What a Coincidence!

Bernhard Wessling

What a Coincidence!

On Unpredictability, Complexity and the
Nature of Time

 Springer

Bernhard Wessling
BWITB
Jersbek, Germany

ISBN 978-3-658-40670-7   ISBN 978-3-658-40671-4   (eBook)
https://doi.org/10.1007/978-3-658-40671-4

Responsible Editor: Eric Blaschke
This Springer imprint is published by the registered company Springer Fachmedien Wiesbaden GmbH, part of Springer Nature.
The registered company address is: Abraham-Lincoln-Str. 46, 65189 Wiesbaden, Germany

I dedicate this book to my grandchildren and those of my life partner, on behalf of all young people today. I hope it will help them in understanding our dynamic and sometimes chaotic world.

Preface

"How could this have happened again? That's impossible!" We've all uttered, heard, and read such exclamations many times. We consider unlikely events to be practically impossible. Experts calculate probabilities of disasters that we and, above all, the relevant politicians trust. There was theoretically nothing worse than a DBA (=design-basis accident) or "maximum credible accident" for nuclear power plants, but in practice, much more happened in Fukushima than was previously considered the worst case scenario. That was "coincidence" ("there were just too many unpredictable things coming together").

Successful people like to claim that they have planned everything, that they have stuck to their strategy. And yet, they forget how often they have met exactly the right people (whom they didn't previously know) by chance or appeared on the market in exactly the right market situation (which they could not influence at all) with a product (which was not intended for this exact market situation at all). Those who have worked just as hard but without success (if they are self-critical) accuse themselves of all kinds of mistakes, but often fail to notice that they have not met the right people by chance, or that they have appeared on the market too early (when the market situation was not yet suitable) with practically the same product, and therefore failed by accident.

August 2021 marked the thirtieth anniversary of the launch of the *World Wide Web*. Nobody could foresee, nobody could predict how much the WWW would turn the world upside down only a few years later. Not even its inventor Tim Berners-Lee, who only wanted to develop it for the easier exchange of scientific documents. Even the way to this start was a chain of coincidences and chances, as he has described.

Rather, we all experienced even more coincidences in 2020: Most likely, a new coronavirus jumped from animals to humans at a wild animal market in Wuhan (China) in December 2019. Perhaps this happened without anyone noticing it before, somewhere else, because according to current knowledge, the first infection occurred in Italy in November 2019, as was subsequently found out. In January 2020, the founder of the biotechnology company BioNTech learned of this virus. He decided to radically shift the focus of the company from cancer therapy research to the development of a vaccine against Covid-19. In December 2020, that is, after less than a year of work, the vaccine was approved. This research was only possible because the founding couple had come across first indications from basic research more than twelve years earlier that *maybe* novel cancer therapies could be made possible using mRNA—we will go into this in more detail. Because, as oncologists, they had founded a company (BioNTech) for this purpose, happened to find two investors by chance—a pair of brothers—who were willing to take risks and be patient; and because they had developed the mRNA technology almost ready for use by the end of 2019, only still with no approved product. The vaccine was possible, even though the research was actually aimed at a completely different application area. But by chance they recognized this possibility of their technology platform and developed a vaccine in an incredibly short time.

We are at least annoyed by very many coincidences. There may be sudden rain, which was not forecast, and the picnic outside falls into the water. Some coincidences are dangerous, such as a car that suddenly takes us the right of way. Some are even life-threatening, such as the accidental spread of the coronavirus to humans. Other coincidences, however, are welcome: We are often happy about small ones. So the sudden reappearance of the sun, after we had packed the picnic quickly and could now unpack it again after all. We marvel smilingly at the bigger one, such as the coincidence that made it possible for me to get to know my life partner for the first time, with whom I live happily together. We shake our heads in joyful disbelief that our favorite soccer club was able to equalize in the very last second, and that only because an opposing defender's leg by chance sent the ball careening in the right direction. The development of stock prices is unpredictable, and this is by far not the only thing that is unpredictable in economic life. Inventions and discoveries are very often, if not predominantly, the result of chance, coincidence and accidents.

Other coincidences are even decisive for all of our lives: The extinction of the dinosaurs, which made the development of mammals possible. And even more decisive was the formation of the moon several billion years earlier, which stabilizes our Earth's axis and thus makes the world we live in possible. A world that we humans could not write about if, by chance, some decisive mutations had not occurred millions of years ago that shaped our brains the way we find them today.

If we're honest, more accidents and coincidence cases happen, more actually rather improbable events than those that we can plan, predict, forecast or influence. All kinds of things go wrong, although we haven't done anything wrong! And just as many events that please us or advance us happen unexpectedly, or at least at a time when we did not expect them. The newspapers and all other media are full of it.

So let's now assume that accidents, coincidences and chances are *normal*, i. e. nothing unusual, but rather completely typical phenomena with us on earth and thus throughout the entire universe; if that is the case—then we also want to know: Where does this come from, how does it come into our world, why is it normal? To answer this question, I will not argue with probability calculations, statistics or chaos theory, as mathematicians, philosophers or psychologists do. Nor will I work with philosophical considerations or psychological analyses that are to make us understand how we can enjoy, endure or cope with coincidence, chance and accidents. First of all—after a more detailed consideration of coincidences and chances that surround us—, we will look at how it comes about that our world consists of nothing but complex structures. If all the substances in the universe were evenly distributed and nicely homogeneously distributed and mixed, there would be no us, with our complexly structured brain, the nervous and circulatory system, muscles, bones, our complex skin, the sense organs and all our fantastically finely built internal organs. How do such complex structures arise? How does complexity that we observe everywhere in the world, everywhere in the universe, arise?

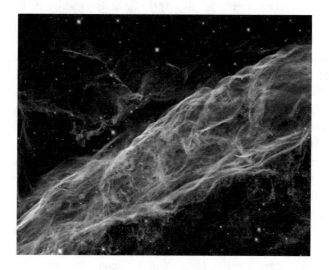

If all substances in the universe were statistically evenly distributed and did not form highly complex structures, we could not admire the beauty of the cirrus nebula. Here is a new picture of the Hubble telescope.[1] *The nebula looks like a turbulent whirling smoke trail, but with a diameter of over one hundred light years. We will look at such and similar phenomena in the book and come to understand the reasons of their formation.*

To avoid making things too complicated, we will also turn to seemingly simpler structures and questions: Why is mayonnaise so stiff, even though the main ingredients—raw egg yolk, water and oil—are each nowhere near as viscous? Why do you have to follow certain steps when making mayonnaise or sauce Béarnaise and can't just pour all the ingredients together into a bowl and stir? Why isn't it easy at all using raw cocoa powder, as used for tiramisu, and milk to make a nice, rich brown cocoa, where cocoa clumps don't sink to the bottom of the cup after a few seconds? How do such complex and constantly changing structures arise as the mouth of a river delta, for example the Lena, as shown on the cover of this book?

---

[1]Cirrus nebula, latest picture of the Hubble telescope; source: https://esahubble.org/images/potw2113a/, Credit: ESA/Hubble & NASA, Z. Levay.

I will try to answer some of these questions based on my own research. For I discovered complex structures in material compositions that could not have existed according to generally accepted ideas. In my research as a chemist, I found a common denominator and therein the cause: All of these products are non-equilibrium systems. It was necessary for me to break away from the idea that I had developed as a chemistry student based on lectures and textbooks, just as all my fellow students had: For us, the world consisted of equilibrium systems, *non-equilibrium* was something rare and unwanted, and thus we described such systems only approximately again as equilibrium systems, ones that were *only somewhat* outside of equilibrium.

In my further research, I learned through the study of the work of many other researchers that non-equilibrium systems are structured. I started to wonder: Why is that so? Or, thinking the other way around: Why are complexly structured systems not in equilibrium? Professor Ilya Prigogine received the Nobel Prize for explaining this with a new theory of thermodynamics. It is simply and solely a matter of entropy, which is very decisive for what determines the course of the world. You may have already heard the word *entropy* before, but either didn't consider it important or possibly not understood it in the first place. I will try to make you understand it. Because anyone who wants to understand the world at least a little from the bottom up should have a reasonably accurate idea of entropy. But don't worry, I will explain it in a way that is really easy and practical to grasp.

Then we will also deal with fundamental questions: If everything was chaotic at the Big Bang, why can order arise in the universe, for example the diversely structured galaxies with countless sun/planet systems? And why can a glowing hot earth become a life-friendly blue planet? How did the Big Bang come about in the first place, wasn't that also a matter of chance or coincidence?

The phenomenon of *chance and coincidence* has always fascinated me. When I was working on non-equilibrium thermodynamics, I eventually came across a connection between chance/coincidence and non-equilibrium at some point. But how can that be? Well, that's exactly what this book is about, and I'll explain it step by step in the course of this book. Suffice it to say for now that both phenomena are inextricably linked to each other. And both phenomena are just as inextricably linked to entropy. My only surprise came, during even deeper research for this book, at the fact that apparently no one had come up with similar ideas before. Or, if someone had had similar thoughts, they had not been written down in a publicly (or at least easily) accessible way. In any case, all I kept reading and hearing was: It is the quanta, the unpredictably behaving elementary particles, that should cause chance, coincidental and accidental events in our macroscopic world. I will explain why this cannot be the case. Quite apart from the fact that there is no verifiable evidence to.

And *time*? We're not going to link entropy to *time* as well, are we? Yes, *we* are not going to link *time* to entropy and non-equilibrium, *we* don't have to and couldn't anyway. Because *time* is already linked with entropy.

But isn't time just an illusion, as Einstein said? And as some other very serious physicists and philosophers also think? But if it is not an illusion, what is the nature of time? We will approach the answer to this question, which has so far been clarified neither by philosophy nor the natural sciences, step by step, just as I have done in the course of my research.

And finally, the older ones among us may also expect an answer to the question (which the younger ones can put it aside for later): Why does time go by faster as we age? Is that really the case, or do we just feel that way? Why does time sometimes seem to go slow for the younger among us, like when we're waiting for something urgent, and sometimes seem to go fast, like when vacation is over sooner than we thought?

If you've ever asked yourself these or similar questions, I invite you on an expedition into landscapes of science that you have probably not visited before. We span wide arcs from simple mayonnaise to complex galaxy clusters, from mundane traffic jams on the highway to the fascinating subject of evolution, from the misunderstood Big Bang to the amazing self-organization of order out of chaos, from surprising goals in overtime to unimaginably oversized black holes. And we thus encompass the becoming and passing in our world, entropy, chance, coincidence and time.

We will also occasionally reflect on how our own thinking might work. How open are we to questioning widely accepted explanations, which we ourselves find pleasant and plausible, about the interconnections in our world? If we are all honest: Most of the time we are not very open. But we should try it more often. In the words of Nobel laureate Daniel Kahneman: "Our mind usually works so that we have intuitive feelings and opinions about almost everything we encounter. [...] Regardless of whether we formulate them explicitly or not, we often have answers to questions that we do not fully understand, and we rely on clues that we can neither explain nor defend."[2] In this book you will encounter some thoughts that you may never have heard or read before. This will make you sceptical, because, as Daniel Kahneman explains in his profound book *Thinking, Fast and Slow* among many other aspects of our thinking, we hold statements that we have read or heard many times before as being far more likely to be true than those that we encounter for the first time. It then requires considerably more mental effort, and

---

[2]Daniel Kahneman (Nobel laureate 2002), "Schnelles Denken, langsames Denken", German edition Siedler-Verlag (Penguin Randomhouse) 2012, p. 127, retranslated from the German text by B. Wessling. (Original English edition: "Thinking, Fast and Slow", Penguin Books 2022). It is particularly interesting to consider his description of two different thinking systems and the experimental evidence for this: System 1 is fast and superficial, constantly trying to construct explanations from all sensory impressions that correspond to previous experiences; it is only, if this does not succeed, that System 2 becomes active, a system that represents our conscious (after-) thinking, but it is "lazy" because it uses a lot of energy. System 1 invents obvious explanations that we then accept. It is happy to answer questions that replace the actual questions, because they are much easier. Then, it seems to us as if the acute question has been answered, which is not the case.

thus more energy (!), to deal with new thoughts and phenomena than to read and hear things that we already know and have long considered to be correct.

With this book, I would like to motivate you to subject your previous world view to a critical examination in part: Is everything really *in equilibrium,* should the climate, the ecosystem, our economy really be *in equilibrium* at best? I do not expect you to change or even overturn your previous ideas; but I would like to encourage you to allow for some unusual thoughts and to think them through calmly. Because only in this way we will be able not only to answer the question of how chance, coincidence and accidents come into our world. We will understand the world as a whole a little better.

Jersbek   Bernhard Wessling

For this English edition, readers should be aware the original German edition is using the German word "Zufall". In English, there are several different words for what Germans are all calling "Zufall": coincidence, chance, accident, randomness. I have tried to use the appropriate term in each given circumstance.

I Thank this World for its Non-Equilibria with the Coincidences and Chances

The origin of this book lies in my unbridled amazement at this world. And about what I found out in my research. I started writing in 2016, having, at the time, just been together with my new life partner for a year. How wonderfully unpredictable my life has been: A big thank you to non-equilibrium of the world I live in, and the coincidences and chances that brought me together with her! She viewed me indulgently when I hacked away at my laptop for hours at the table in our living room-kitchen in our first small flat together in Germany or at the sofa table in China. And not only there, but also on the plane to or during our trips within China. For two years. I am very grateful to her, above all for tolerating and supporting my work on this book—in parallel to my other books[3]. But even more than that: She had read the pre-pre-pre-version during its creation in 2016/17. But then stated: "I won't read the book again until it's really finished." So, she knew, even before I did, that it would take many years and many revisions and fundamental changes until the book was *really finished*, assuming it ever really was. It indeed got finished. She was surprised how much had changed when she read the original German edition.

Of course, not all the people to whom I am or should be grateful can be found here—the list would be too long for this book. But those not named can be sure of my gratitude! Nevertheless, I would like to mention a few key people.

Without Professor Ilya Prigogine and his scientific life's work, I would never have found the solution to the puzzles that troubled me so much for years, if not decades. Professors Werner Ebeling and Grégoire Nicolis specifically pushed me towards an essential question that I had to work on and could solve. The theoretical physicist Dr. Helmut Baumert let me lure him into a question that was initially uninteresting to him, until we were able to jointly develop a new theory of turbulence in non-Newtonian fluids. The non-equilibrium thermodynamicist Dr. Rainer Feistel,[4] the physicist Prof. Serdar Sariciftci

---

[3]"Der Sprung ins kalte Wasser", Verlagsgruppe Eulenspiegel 2023 sowie "Der Ruf der Kraniche", Goldmann 2020/"The Call of the Cranes", Springer Nature 2022.

[4]https://www.io-warnemuende.de/rainer-feistel.html

(Johannes Kepler Universität Linz, Austria)[5] as well as two other scientists read an earlier version of the manuscript and made a lot of critical comments. These have prompted me to do additional research and to think even more deeply, which is reflected in the book. My younger brother gave me numerous hints for passages that, despite my best efforts, had not been written in a manner that would be understandable to everyone. But, in particular, without the work of my many employees, especially in the laboratories of my company, there would have been nothing for me to discuss with these outstanding scientists. I have worked very closely with many of my employees, some for more than twenty to thirty years. The extremely intense thirteen years I spent in China also fill me with gratitude. I cultivated friendships there, that are still alive today. Our common active time was full to bursting coincidences, chances, accidents and exciting scientific discoveries ...[6]

The freelance editors Obst & Ohlerich criticized an early version of this book to the ground in 2017, but they found the final version convincing. I learned a lot from the final round of editing carried out by Rouven Obst of the original German edition. Now, this English edition was reviewed by Marc Beschler (who already had reviewed the translation of my previous book about my sideline crane behaviour research, "The Call of the Cranes" (SpringerNature 2022), and I am again so grateful to him. The German poem in the Final Remarks (at the end of the book) was translated by Timothy Adès, a "rhyming translator-poet", whom I found "by chance" (sure! how else could it be!) and with whom I exchanged a few emails until the final version was approved by both of us, we both had fun. Let is be known that I am solely responsible or all remaining errors. In his editing of the original German edition, Eric Blaschke (editor at Springer Vieweg) pointed out numerous minor and major deficiencies, which prompted further research and improvements to the text. The publisher Springer Nature, here in particular the responsible editor Eric Blaschke and his boss Dr. Garbers, took the risk of publishing this book project, and now also this English edition—for this I am particularly grateful. Especially because Springer found my book project interesting and worth publishing, after I had already collected rejections from countless publishers and literature agencies before. Actually, I had already given up hope that a publisher would take it and want to bring it out. What a coincidence that I finally asked at the editorial team of Springer Vieweg!

But that is exactly how our world is, constantly things are happening that one cannot imagine, that cannot be foreseen, and for that I am grateful to the world for it! How boring my life would have been, if everything had been predictable and everything that I was to experience in my life had been documented in the appendix to my birth certificate. Fortunately for all of us, however, the world is made up of nothing but non-equilibrium systems—full of grace, beauty and surprises.

---

[5]https://www.jku.at/en/institute-of-physical-chemistry-and-linz-institute-for-organic-solar-cells/team/sariciftci/

[6]https://www.researchgate.net/publication/260427241_Milestones_highlights_of_the_Organic_Metal_Polyaniline_Science_Technology, see Appendix 15.

Contents

Vita

Dr. Bernhard Wessling, born 1951, raised in Herne, studied chemistry in Bochum. From January 1978 to the end of 1980, he worked for an engineering company in Düsseldorf. After that, he moved with his then wife and two small children to Bargteheide and took over the management of the development laboratory of Zipperling Kessler & Co., a company with fewer than fifty employees, but a 170-year history. A few years later, he took over the management, and soon he became a shareholder of the company (with a very risky buy-out of a previous shareholder). It subsequently developed into a successful company with three hundred employees. At the end of 1995, Wessling founded the daughter company Ormecon, selling all of Zipperling's business to Clariant AG mid-1996. Wessling, having relocated in 2005 to China, sold Ormecon at the end of 2008, founded his one-man technology consulting company in ShenZhen and remained in China until the end of 2017. Throughout his career, despite holding such positions as managing director, he continued to be directly responsible for research and development. Thus, even as he fashioned himself as an author and entrepreneur, he remained committed to pursuing basic research.

Chance and Coincidence Take their Course

1

Abstract

The author describes numerous instances of chance and coincidence that opened his way into science, into basic research, in a small company that he has been responsible for as managing director and shareholder at a very early stage. These coincidences led him deeper and deeper into areas that have previously received little scientific attention and where surprising phenomena waited to be discovered.

About 30 years ago, on August 6, 1991, the world's first website went online.[1] The initial ignition of the *World Wide Web* was a product of the European research center CERN—created by accident and as the result of many coincidental events—, albeit initially only an internal one. Two years later, on April 30, 1993, the WWW was activated for the general public. Access was free of charge, but there was considerable cost around the technical requirements and the high telephone charges at that time. And still hardly anyone noticed it, certainly, no one predicted at that time that the WWW would have an impact on practically all areas of our lives. It was intended as a platform for the documentation and international exchange of documents within CERN, and even that came about unplanned.

Only two years later, on October 5, 1995, my company became the second chemical company in the world to go online, the next coincidence. I was and am anything but a

---

[1] Here is a copy of the original website: https://www.w3.org/History/19921103-hypertext/hypertext/WWW/TheProject.html; or here: http://info.cern.ch/hypertext/WWW/TheProject.html.

Supplementary Information The online version contains supplementary material available at https://doi.org/10.1007/978-3-658-40671-4_1.

computer nerd, yet I almost landed on pole position. Only the then-chemical giant Hoe-chst AG had beaten us by just a few days. But our site was much more informative: I had ensured that *all* our technical information and my scientific publications (together, over four hundred documents) were available, while Hoechst offered only the homepage.[2] In1995, in the Internet, which was intolerably slow from today's point of view, housed only 23,500 web addresses.[3] Today (as of Sep 12, 2022) there are two billion,[4] that is, over 80,000 times as many. As to our own site, I had no further ambition back then but to make our technical and scientific information more easily available to our international customers and research partners worldwide back then. There was no notion at all as to what the WWW would become in just a few years to come.

I was only reminded of the fact that the WWW would celebrate its 30th anniversary in 2021 or its actual anniversary as a public Internet in 2023 thanks to the article *How the Web Came Into the World?* at the end of 2020 from the weekly magazine DIE ZEIT.[5] And although the platform is therefore, strictly speaking, still very young, it seems to me as if the Internet has always existed. My memories of the telex, which I myself used actively for many years, as well as the fax machine that followed, strike even me now as being akin to stories from the time of early industrialization.

How many coincidences led to the existence of the WWW? Above all, there was never a plan to build something that even remotely resembles the Internet as we know and use it today. Tim Berners-Lee, who is considered the inventor of the WWW, once said that there was never *the one* decisive idea that initiated the Web: "I wrote the first programs. But many other people have contributed important ingredients, just as randomly as I did." Bernd Pollermann, who also worked on this chaotic team, said: "Why did I join in? So that my colleagues would no longer annoy me with their never-ending requests for internal telephone numbers." So, he just wanted to create a telephone directory that was accessible to everyone and could be edited easily, nothing more. Neither was the project an official project of CERN, where the developers were employed, nor was anything more planned than easier availability of scientific documents and telephone numbers of this huge European research institution; such information was previously almost impossible to find, even for insiders. The Englishman Berners-Lee had already completed an internship at CERN in 1980, after which worked for four years at

---

[2] In this web archive, you will find a copy of the websites from back then, not directly from October 1995, but from November 1996, because in 1995, there were also no web archivists: https://web.archive.org/web/19961125121325/http://www.zipperling.de/ZKC/About (with the hyperlinks that also work in this web archive, the reader can call up other websites from my company from that time); however, only the very few initial Hoechst websites from January 1998 can be found there.

[3] "unique hostnames"

[4] http://bit.ly/3HXSli1 "How many Web Sites are there?" (Sep 2022)

[5] https://www.zeit.de/2021/01/word-wide-web-erfindung-internet-cern-genf-physiker-forschung-szentrum.

a software company before applying and being accepted for a job at CERN in 1984. In 1989, he proposed a project that aimed to make research results easier to exchange and find. This was a self-imposed task for which he had most assuredly not been hired. His direct superior said: "Vague, but exciting." The project was run unofficially, because it would have taken far too long if a project application had been made. And it probably would have been talked down in the committees. When Berners-Lee wanted to present his concept at a hypertext conference at the end of 1991, he was not allowed to give a talk—even though the CERN-internal WWW already existed! He presented the state of his development in the foyer, where it hardly attracted any attention. When the Web was eventually opened up to the public in 1993, Berners-Lee was still only concerned with the exchange of information among researchers.

I too merely thought of providing technical information, albeit half already to customers, only the other half to my research partners and research competitors. A seemingly endless chain of chances, accidents and coincidences had led me to a point at which my advertising consultant and I began discussing, in the spring of 1995, how we wanted to present ourselves in the following autumn at the biannial plastics fair in Düsseldorf, the world's largest fair of its kind, which takes place every two years. As always before this event, we were looking for a funny or attention-grabbing action with which we could attract notice. One possibility was to advertise our active involvement in the Christo couple's wrapping of the Reichstag, the cover of the invitation to meet us at the fair is shown in the lower left-hand image. We had developed the flame retardant concentrate for the wrapping film and ropes to be used in Christo's project (see also Appendix 4).[6] But it was still unclear whether the project would go on as planned, because the wrapping was not due to begin until June and the approval was overdue. For the fair, it was still very uncertain whether we would get the two Christo lithographs (shown in the lower middle and right-hand images) that we had been promised for a raffle—a plan we developed for the show, which was a prerequisite for an effective advertising campaign.

And so, this other idea wafted around the room: Two years earlier, the *World Wide Web* had become generally available, and there had been first tentative attempts to use it for company information. I had already heard about this on the sidelines, while my consultant had read a lot more about it and had almost developed into an expert. At that time,

---

[6] https://web.archive.org/web/19970301045342/http://www.zipperling.de/News/Christo/reichsta

not a single chemical company in the world was online with a website, let alone one from the plastics industry, my company's particular market. I thought we could try to be the first. So, we got started. This endeavour turned out to be an obstacle course, with a dozen of my employees, my consultant and I putting in a lot of hard work. Ultimately, we managed to get everything ready a mere few hours before the fair's opening that our homepage went live. And so, we killed two birds with one stone: We made the relevant research results from our house available worldwide and, at the same time, created a stir.

Our company's booth at this world plastics fair, *K'95,* was constantly surrounded by a large crowd of people, especially since we were also raffling off our Christo lithographs. The visitors had, at best, heard or read about the WWW, but had never seen anything like it themselves; now they wanted to see it. Nobody had a computer at home with which they could have accessed the Internet with a beeping modem; only a few companies had something like that. But we had set up the then-modern Pentium computer with a large tube screen, connected it to the WWW by telephone line, and my employees and I took turns demonstrating the new websites and their content. The browser available at that time was *Netscape*, and if one wished to do a web search, one turned to *AltaVista.* We even printed out whatever website pages interested visitors desired on request. It was *the* talk of the day at the fair.

We were something like small pioneers on the WWW at that time. And at the same time, we were active with our insignificant company in what was then a *very hot* area of basic research and that would later shape my professional and business future. The road there was paved with an incredible number of chances, accidental events and coincidences, just like the road to the WWW and many other inventions, discoveries and important social changes. Above all, my scientific path showed me the deep cause of the perfectly normal occurrence of coincidences. But let's take a few steps back to the first accidental event without which I would not be alive, before we continue with the professional coincidences.

Many people probably tend to say that their birth and their conception before that were rather lucky coincidences. In principle, however, this applies to all people. Because only a fraction of the billions and billions of intimate encounters between two loving people of different sexes actually leads to inception. In my case, however, it was evident that my parents did not want to have a fifth child, at least not at *that* time, because their fourth one had just been born. But two months later, my mother was already pregnant again, namely, with me. I was born only eleven months after my next-older brother, and only six years after the end of the war. My father was still unemployed, our family was downright poor. A reasonable married couple could not want to have another child at that time. Either my parents were unreasonable at that time, or they simply had bad luck with contraception. But if you wanted to use contraception as a Catholic in those days and not be condemned to hell at the same time due to a mortal sin, you had to plan sexual intercourse according to Knaus-Ogino[7] or, even better, with a thermometer, which often helped the Catholic Church and the German people to gain more members. On this occasion, by accident, that new member was me. (My mother has later confirmed this to me.)

As a child and teenager, I was curious, adventurous, and fond of exploring, and I took an interest early on in scientific questions; I loved biology class, where I learned about chromosomes and DNA. Watson and Crick, who had elucidated the structure of DNA,[8] became my idols. It wasn't long before a firm decision had formed in my head: I wanted to become a biochemist. Here, the next chain of coincidences began: I actually found myself very enthused with the idea of going to Tübingen (my oldest brother was a physics student there), the town's university was, next to Hanover, one of only two in Germany where you could study biochemistry. As a boy born in Herne in the middle of the Ruhr region, however, I also looked at the newly founded Bochum Ruhr University not far away, for cost reasons, and because I didn't want to be too far away from my first girlfriend. I went there to visit the chemistry department's biochemistry professor, who had come from the University of Tübingen. He advised me: "Don't study biochemistry. You would then be neither a biologist nor a chemist, without the ability to do either of them right. Study chemistry and also some biology on the side, then you can later decide which path in chemistry you want to take." That's what I did.

Unfortunately, the completion of my dissertation on a complicated synthesis question in natural product chemistry happened to coincide with a major economic crisis: The entire European chemical industry had imposed a hiring freeze. After a period coinciding with my high school years during which the industry had tried desperately to recruit chemists, I was now, after only seven years, a graduated chemist at a time when they weren't hiring anyone anymore. I decided to go to a small three-man engineering consulting firm, against the advice of everyone around me. I had seen a notice on the bulletin board of the chemistry department and applied.

My laboratory neighbors blatantly tapped their foreheads. "Are you crazy? You have to go either to a famous university or to one of the world's leading chemical companies and do research on big projects there. You're as active and efficient and goal-oriented here as no one else—and now you want to stop engaging in research?" But who said I wanted to stop researching? Maybe I could get something going on my own in the young engineering company? "That will never work," my best friends replied.

---

[7] Knaus and Ogino had been the ones who independently developed a calender-based contraception method which was recommended by the Catholic Church and usually referred to as "Knaus-Ogino method", cf. also https://en.wikipedia.org/wiki/Calendar-based_contraceptive_methods.

[8] J. Watson, F. Crick, Nature 171, 737–738 (1953), link to original paper: https://www.nature.com/scitable/content/Molecular-Structure-of-Nucleic-Acids-16331.

From Plow Horse to Racehorse

塞翁失马 (Sài Wēng shī mǎ),[9] very freely translated: "Who knows what will develop from it" or "You can never know". I only know this proverb since I lived, worked and also learned Chinese in China for thirteen years, from 2005 to the end of 2017, as a result of further accidents and coincidences. It is one of the infinitely many Chinese 成语, the *Cheng Yu*. These are proverbs that mostly consist of four characters, behind which lie stories that are thousands of years old. They contain valuable wisdom that Chinese students still learn today and—if you learn to understand them and really digest them—can be very helpful in coping with and understanding life.

塞翁失马 (sài wēng shī mǎ, "Sai Weng loses a horse") is the condensed, four-character[10] comprehensive headline of the following little story:[11]

> There was a man who lived on the northern border of China, whom the people called Sài Wēng (塞翁; old border dweller). One day, his horse broke loose and galloped into the barbarian territory beyond the Great Wall. The neighbors visited Sài Wēng to express their sympathy, but Sài Wēng surprised them by asking: *"Who knows if this is not a blessing?"*
>
> Months later, the horse returned with a mare, an incredibly beautiful and very fast horse. Sài Wēng's possessions had more than doubled overnight. Many came by to admire the new horse and congratulate him, but again Sài Wēng showed no great emotions. He said: *"Who says this can't be bad luck?"*
>
> Sài Wēng's son proudly rode the new horse around, but couldn't hold it well, fell and suffered a compound fracture of the leg. Again, sympathetic neighbors came, and again Sài Wēng said, as calm as ever: *"Who says that can't be some kind of blessing?"*
>
> A year later, the barbarians crossed the border. All young men fit for military service were drafted into the army to defend the country. There were heavy casualties on both sides.

---

[9] The Chinese characters are made "readable" for foreigners by the official romanization "PinYin" with accents, which indicate the emphasis of the syllable that one has to use when communicating with Chinese people.

[10] In full, this proverb is: 塞翁失马, 焉知非福 (Sài Wēng shī mǎ, yān zhī fēi fú), thus consisting of two times four characters, literally translated as "Sài Weng loses a horse, how do we know not a blessing?" In German, this is often translated as "luck in misfortune". But the Chinese Cheng Yu only partly corresponds to our German saying, because what we mean by that is: When something unfortunate happens, for example, an accident in which one only sustained a superficial wound, despite the fact that it could have been a broken leg or even worse, one had luck in misfortune. The Chinese one goes much further, and means: Even if something bad happens, something good can develop from it (and vice versa). The accents on the syllables indicate the emphasis, which is decisive for communication in Chinese as a tonal language.

[11] I partly took the text from https://ostasieninstitut.com/enzyklopaedie/saios-pferd-%e5%a1%9e %e7%bf%81%e3%81%8c%e9%a6%ac-saio-ga-uma/, revised and translated into English. Reprint with kind permission of the institute.

Nine out of ten men died. But Sài Wēng's son was not drafted because his leg was crippled. Therefore, he was spared the terrible fate and his family survived the war completely.

So, at that time, I had decided to set my sights low, very low. I had consciously decided *not to* pursue a university career as an alternative to big industrial chemistry. I considered myself incompatible with the research philosophy and bureaucracy of the universities. And apart from that the chemical industry had no openings at all, I was also not interested in applying for a position at any of the giant chemical companies where I was afraid I would be drowned in their hierarchy. So I started as a lone fighter and "chief" of a one-man "chemistry" department, handled consulting contracts for which some little chemical understanding was helpful, got new projects and actively sought opportunities to start solid development work without any realistic prospect of real research. By chance, I came into contact with an older engineer who had a strange idea: to "stretch" plastics, so to speak, by incorporating fillers, thus to dilute them and making them cheaper. I didn't have the slightest idea about plastics, but I set to work, practically alone.

鹰单飞，羊群集。 (Yīng dān fēi, yáng qún jí.), "Eagles fly alone, sheep flock together". Another interesting Cheng Yu, but at the time I didn't feel like an eagle at all, but rather like a lonely gray mouse.

So, by chance, I was led to a question in the field of plastic technology. Through a combination of incredibly improvised means, hours spent working on rented equipment in the halls of machine manufacturers and with advice from this senior engineer, I succeeded in developing a new process: Using vacuum in mixers and extruders, I worked a considerable amount of mineral filler into conventional plastics without significantly compromising the positive properties of these materials.

After my Ph.D., I had lost my horse—the prospect of an attractive job in chemical research—and unlike Sai Weng I had not found another particularly fast horse, but rather a lame donkey, namely, the position at the engineering consulting company. But I made a useful workhorse out of the lame donkey: I developed a new process, filed a patent,[12] won an investor, built a small factory and started my first combination of technical development with marketing and management. At the same time, the oil crisis developed worldwide—what a coincidence! Oil, and thus also plastic, were suddenly scarce and expensive. Without anyone—let alone me—having been able to foresee it: Our products were suddenly in demand!

In the meantime, a friend of mine, a laboratory neighbor from the time of my doctoral thesis, got in touch. He was doing postdoc research in France on electrically conductive complex salts. We were in contact by letter, and thus he knew that I was on my way to working on plastic technology. Now, he sent me an article that had just been published, in which three professors and two of their employees reported the discovery of

---

[12] https://patents.google.com/patent/WO1981003144A1/en.

so-called electrically conductive polymers.[13] This electrified me; I began to think about how I might establish this as a research project for my small company in the future. I developed first ideas. It was clear to me that it would take a long time until *conductive polymers* (the behaviour of which is the smooth opposite of that of normal polymers, i.e. plastics, which are insulators by nature) could be used industrially. Maybe I could contribute something to it? My workhorse suddenly seemed to be a racehorse, with many interesting prospects being offered by pure chance. But the young German Sai Weng had to experience that he fell off this horse before it even started running:

The three consulting company founders, for whom I was an employee, decided that I was miscast as a chemist for plant management and process/product development. And that, even though I had developed the product and the process for its production almost completely alone, I had also designed the factory, managed and been responsible for every aspect of its construction and operated it with my new employees, all on my own. The engineers were of the opinion that I had done very well in marketing and as a sales-person, because there had already been first and significant sales, and there were many more potential customers. Now, this small and successful company, which I had previously led alone, had to be led by engineering professionals like themselves. I, as a chemist, should concentrate on marketing and sales. This was astonishing and much more than a little short-sighted, because in all previous instances, they hadn't even remotely cared about the process for one single minute, didn't know anything about it, and had no expertise in it. So I rejected their request and gave them three months to change their minds. In parallel, I was looking for a new job in case they let my ultimatum expire.[14]

So I found myself—without knowing it at the time—at a very decisive rossroads. At which point the absolutely most unlikely coincidence of all the coincidences that have happened to me in my life occurred. In my distress, I had placed an ad in a Wednesday edition of the "Frankfurter Allgemeine Zeitung", FAZ. The *Stellengesuche Akademiker* (academic job postings) back then occupied about half a page in the weekly science section. My ad only took up one column and was only three centimeters high. Tiny. Wolfgang C. Petersen, managing director and 40% shareholder of Zipperling Kessler & Co.[15], was, of course, a subscriber to the FAZ, but *never* read the science section, let alone the academic job postings there. But on that Wednesday night, he leafed through this part of

---

[13] H. Shirakawa, E.J. Louis, A.G. MacDiarmid, C.K. Chiang and A.J. Heeger, J Chem Soc Chem Comm (1977) 579, for which they received the Nobel Prize in 2000.

[14] I was actually interested in staying, because, on the one hand, my then -wife had just given birth to our second son, and, on the other hand, I was in the process of renovating a rundown post-war house on a dead end street on the outskirts of Düsseldorf, only 100 m from the edge of an extensive forest. But despite any financial advantages this change might have had for my situation at that time, just having bought the house, just having got a second child, I still could not endure being downgraded to a mere salesperson.

[15] https://www.researchgate.net/publication/262076086_Short_History_of_Zipperling_Kessler_Co_Ormecon's_mother_company.

the paper for the first time in his life, as he was plagued by business worries and couldn't fall asleep. His bored gaze fell on the ads and he thought: This is exactly the man we need. That very night he made a note for his secretary to get in touch with me.

When I later discussed such stories and topics with my Chinese friends, it was clear to many of them: This is "fate", 命运 (mìng yùn), everything is predestined and inevitable, we just don't know it. But I insisted just as stubbornly that, no, it's just a coincidence, that's how the world works.

To stay with the story of the old poor border farmer Sai Weng, I, the young Sai Weng from the Ruhr region, had exchanged the lame but useful plow horse (which only briefly presented itself as a racehorse to me) for an old but apparently solid, confidence-inspiring horse. I had become the head of a small lab with the task of developing new products for the company. Zipperling was still dependent on so-called record molding compounds for well over ninety percent of their business, even though they had had to make mass layoffs just two years earlier as a result of the success of the *compact cassettes*. The plant had shrunk from a proud one hundred and forty employees to less than fifty. And yet, there were only minimal approaches to or ideas for new products and markets.

Such a task corresponded to the ideas I had developed for myself towards the end of my doctorate. Already as a doctoral student, I had buried my dream of becoming a successful researcher, an internationally recognized university professor who makes world-changing discoveries. It was clear to me: I was no genius. But here at Zipperling, I could be responsible for a small, independent area, even if it was only with two technicians directly reporting to me, and develop it. I would immediately feel the impact of what I was contributing, whether positive or negative, productive or counterproductive. So, the three years at the consulting company had become a test run for the work I could now resume in a laboratory, work from which I might be able to develop real research. And so, I started dreaming again.

But once more, the German Sai Weng fell from the apparently solid horse before the first steps could be taken: On a Saturday early in November 1980, my young family and I moved to Bargteheide. We had the employees of the moving company unload the truck, distributed everything in the house and then sat down to take a deep breath. The children (Boerge, one year old, Bengt, two and three quarters) were happily playing in the chaos. The doorbell rang. At the door stood Wolfgang C. Petersen and his wife, with a large, round cake, which turned out to be a huge Lübeck marzipan cake, the usual Christmas present from Zipperling for their customers. "May we come in for a moment?" they asked politely, because they had come unannounced.

My unspoken question as to why they were already visiting on the day of our arrival was quickly answered: "Dr. Wessling, we have huge problems at Zipperling. All of our record molding compounds are suddenly causing scratches on the matrices. And our blown film colour concentrates have been rejected, everything we've produced for the past few weeks is leading to hideous smelling plastic bags. But the worst thing is that the smell only starts when the film has already been processed into bags by the customers of our customer."

I had no idea what he was talking about. Record molding compounds and scratches? Color concentrates and stench from plastic bags? Where had I gotten myself into? The moving boxes were all still unpacked, one after the other enthusiastically inspected by my small sons. My new job wasn't supposed to start for another two months; why did I have to talk to my future boss in the midst of all this mess? To talk about complaints that I didn't understand and that I assumed were not part of my job description? "If you can't solve these problems, we'll go bankrupt soon," was Mr. Petersen's prompt response to what was presumably my extremely confused look and my heavily furrowed brow. "Because my employees can't do it, as I've noticed for weeks."

I actually felt as if I had been thrown off the horse in a high arc and kicked by it afterwards. Finally, I had found a position that had raised my expectations, and there I was, already confronted with an existential dilemma. Either I solved the problems, or I was unemployed before I had even really started the new job. I had no choice: I agreed to come to the company directly after the weekend on Monday morning.

With melancholy, as well as thoughts of my family and the many moving boxes sitting around unpacked, I set out, began with a deep chemical and physical analysis against much resistance (mainly from the plant manager and from lab technicians), and then moving on to using modern methods and external equipment. I found the causes within about three months and eliminated them (see also Appendix 1). This intensive work led me deep into the world of plastic compositions. In this short time, I made contacts with external laboratories, developed analysis methods, got to know production methods and problems, and developed a concept for my laboratory—all necessary conditions for the later successful work. Without this enormous existential pressure, the learning curve would certainly not have been nearly as steep.

Sai Weng from the Ruhr region therefore now believed that he was firmly sitting in the saddle of the solid workhorse. Already years ago, I had reduced my expectations of the future. I did not want to drown in the intrigues of the academic world or in the hierarchy of the big industry and so ended up in a three-man laboratory of a plastic compounding company with around forty employees in Schleswig-Holstein. I had not just *resigned* myself to that, but it corresponded to my ideas. And the bankruptcy risk was also over. But for exactly sayable six weeks.

In March, Mr. Petersen had to apply for short-time work. While my small laboratory team, despite short-time work[16], worked overtime—the only exception, to the protest of the employees—in order to finalize some product development, a completely new

---

[16] In Germany, there is a law which allows companies under certain conditions to apply for financial support in case of critical lack of orders; if agreed by the authorities, the labor office will pay (out of unemployment insurance money) a significant amount of the salary of those employees who have to work short time, e.g. only 50 % of the normal time. That would mean, the company only pays 50 % of the salary, the unemployment insurance pays 40 %, so that the employees get 90 % of their former salary. See also https://en.wikipedia.org/wiki/Kurzarbeit.

request from a large chemical company arrived: In one of their many factories, they had converted facilities for the production of flame-retardant foam panels from powder processing to the use of concentrate pellets (a so-called "masterbatch") at a cost of millions. But they could not start the new production because no one could supply them with the concentrate that they had developed on only a small scale. They had not considered tests on a large scale to be necessary, but when their suppliers tried to produce it on a tonnage scale, their large facilities failed, the product decomposed. The chemical giant's stock emptied at a alarming rate, until they finally came to us and asked us to solve the problem which I managed to do within a very short time. In this case, too, I was able to solve the problem based on a clean problem analysis, followed by a creative chemical idea (Appendix 2). Just in time. Because the product was one of the most important end products of this chemical giant, it went directly to the consumers via the wholesaler.

This request came equally just in time for Zipperling, at literally the last moment. With hundreds of tons produced per year, it became one of our company's biggest product. What a coincidence.

The chemical company had probably taken so much time to ask us for help because Zipperling was simply unknown and, frankly speaking, technologically incompetent. This incompetence also extended to marketing: Mr. Petersen did not like to spend money on advertising, and what he did spend was essentially money down the drain, used for maybe five-centimeter-high, meaningless ads in the style of the 1920s. I was horrified when I saw them.

But again, by chance, help was in sight: I had bought a small post-war house in Düsseldorf that I was not able to sell a year later when I moved to Bargteheide because the real estate market had collapsed. So, it had actually been possible for me to eclipse Sai Weng. I had therefore first rented out the house, which I had finished renovating just before moving out. Now, I remembered that my Düsseldorf tenant had only recently founded an advertising agency. I spoke to him during a visit and arranged a later meeting with him in Ahrensburg. I suspected that there was an untapped market for specialty products, of the kind we had developed for this large chemical company. My new business idea for Zipperling was to offer such products that other compounders (our large or small competitors) did not want to or were not able to develop. The experiences with our new major customer motivated me to do this.

With my Düsseldorf tenant, I designed a marketing concept tailored to this under the motto: "Make your innovation our job." So, if a customer wanted to develop a new product, a new process that required a special product, or, even better, a product complex to develop and to produce, they should ask us. No matter what thermoplastic was involved. The advertising agency designed a protagonist for this, drawn with no more than a few strokes: the caricature of a somewhat prematurely-aged professor, dressed in a white lab coat. In the following months and years, this Mr. Professor presented our latest innovations and solutions for customers in advertisements and brochures in different variations.

My then-still small children, whom I often took with me to the laboratory when I had something to do there on weekends, said: "Dad, that's you!" Well, the only similarity was actually the lab coat.

And 14 years later, at the international plastics trade fair K'95, we presented the world's first truly informative website of a chemical company. Thus a chain of accidents and coincidences and all kinds of forks in the road completed the circle. They opened up deeper and deeper paths into the unknown landscapes of science, which we will look at step by step in the following chapters.

A few weeks later, I founded a new company as a subsidiary of Zipperling: Ormecon. I wanted to try to market my now almost market-ready *electrically conductive polymer*, which I will tell you more about in Chap. 3. I had long since become managing director and shareholder. Zipperling had meanwhile grown to three hundred employees, producing one and a half thousand different special products for hundreds of customers every year, representing around 25,000 metric tons per year. We were experiencing an enormous growth. I was aware that we would not be able to successfully plough two such different, non-overlapping business areas as a small company. We also lacked the capital for this. So, I looked for venture capital investors.

Again, help arrived by chance: Only a few months after the fair, I received a call from a member of the board of Clariant AG, a world-leading and very important specialty chemicals company. Clariant was also one of our competitors. The board member, whom

I had previously met at a meeting, had got wind of my search and, without much pre-amble, came bluntly to the point: "We can provide the capital if you sell the Zipperling business to us." We agreed on a meeting, which took place only a few days after the phone call, and three months later, we signed the purchase agreement. This had all come about despite the fact that I had already rejected a takeover offer five years earlier. Yes, what a lucky coincidence!

More details on this chapter in appendices 1 to 4.

Essential Uncertainty is Everywhere

2

Abstract

Based on numerous examples from recent times (including the Corona pandemic), scientific research, technology, economy, history and politics, one can clearly see: It is the so-called *essential chance* that determines the course of history, and *essential uncertainty* can be found in all aspects of life and nature, especially evolution that our uncertainty about future events is essential. We will also deal with explanations of the causes of chance, coincidence, accidental events and randomness that have been presented so far by other authors from physics and philosophy. The *essential chance* (or: *essential coincidence*) is to be distinguished from this.

It could be that part of my readership may come to think that now really too many coincidences had been mentioned for them all to have really happened. That this might clearly be a case of a failed, as well as dilettantish, novelist going gaga. No, I guarantee that none of this is made up. I would simply like to encourage you to ask yourself attentively and critically whether it doesn't work out similarly for all of us in life. Some could now object that some people have a lot of good luck and others have a lot of bad luck (both of which are chance, coincidences or accidental events). For me, as is probably the case for many other people today, the range of possibilities for events in life has increased enormously, and we are also now capable of learning more about the whole world through all kinds of media. That's probably why we experience more coincidences and hear or read about many others.

But there is one statement that most Germans (with the exception of a large faction located mainly in the south of our country) would likely agree with: Bayern Munich as a Bundesliga soccer club has more luck than other teams, with coincidence most often channeling its good luck into the feet of the Bavarians. For example, on May 19, 2001, in Schalke stadium, when Schalke 04 was already feeling having become the German soccer champion after the last minute of the last game of the season 2000/2001. But then, everything changed in the game in Hamburg that was running at the same time but had not yet finished, a game that Bayern Munich would have to lose in order for Schalke to receive the championship trophy, and when the game in Schalke had finished, Hamburg's soccer club was leading: But in the very last minute, something went wrong, Bayern tied the score—the final whistle blew, and Bayern were champions, while Schalke were only "champions at heart".[1]

But whether Bayern Munich experiences more coincidences that lead them down the winner's or the loser's road is probably more a question of perspective. For example, in the final of the Champions League 1999 against Manchester United in Barcelona. There, the Bavarians led 1–0 until the 90th minute, when in overtime the Englishmen scored twice, bringing the score first to 1–1 and then to 1–2. The tying goal was definitely a product of luck, because the long pass forward was a blind shot of desperation.[2]

In soccer, most people would probably admit: it is perhaps normal that coincidence and accidental events so often prevent or produce a goal. But not otherwise in life! However, let's be honest: If you think intensively, talk to parents, siblings, other relatives and close friends, you will also find that such chains of coincidences have also affected you yourself, as they have significantly influenced the fate of each of us.

An older acquaintance recently told me about her marriage: In the first year, it crashed so often that she filed for divorce at the end of the year. However, then the lawyer's letter was subsequently lost, and thus the matter came to nothing. And lo and behold, the marriage became a happy one. As I write this paragraph, the couple has been married for 48 years and is happy that the letter got lost.

Sometimes, it was a teacher who gave a special advice or encouragement, which then led to the choice of a fulfilling profession or an important hobby. Maybe it was just the decision by childs' parents to move to another town, which led to the childs losing friends, but allowed them to enroll a school that promoted their talents and aroused their interests like never before. An accidental bout with a serious illness can, in some cases, lead to a completely different view of one's own life, turning it upside down. Everyone can find plenty of examples of more or less serious coincidences and accidents that caused either a negative or a positive change.

The Accidental Corona Pandemic

We all became witnesses to the consequences of a drastic accidental event at the beginning of 2020, which still has far-reaching consequences. As I became more and more involved with this book in 2019, based on work done in previous years, a virus jumped

---

[1] https://www.sport.de/news/ne2229064/fc-schalke-04-die-schwaerzeste-stunde-der-vereinsge-schichte---fc-bayern-schnappt-titel-2001-weg/

[2] https://www.spiegel.de/geschichte/fc-bayern-muenchen-vs-manchester-united-finale-1999-mut-ter-aller-niederlagen-a-1269035.html

to humans, a new coronavirus, *SARS-Cov-2*. Even now, more than three years later, the exact time of the jump to humans and the source of the virus are unknown. It is very likely that it happened in Wuhan (China) at an animal market, in November or December 2019. It could also have happened somewhere else and earlier before, because, according to information at the beginning of 2021, the earliest documented infection case was an Italian woman living in Italy who had no direct connection to China.[3] There are also serious assumptions,[4] that the virus had been spreading among humans for much longer, but had not previously caused any diseases, or at least no conspicuous symptoms (except perhaps a cold or cough), until, after a series of mutations, not only was it possible for the virus to spread successfully, but also for it to cause serious and sometimes lethal diseases.

However it happened, whether it was through bats or other animals,[5] that the virus could not harm, it does not change anything: It happened by accidence. It is coincidence that it can multiply in humans; coincidence that it makes them sick (most viruses are harmless or even helpful—we also have genes in our genetic material that come from former viruses); coincidence that it is also so highly "virulent", that is, contagious; and, last but not least, it is also coincidence that it is fatal for certain risk groups and people with pre-existing conditions and that, even after recovery, it requires months of convalescence in many cases, some of whom may never really be healthy again, suffering from what we now call "long Covid". This unfortunate combination of coincidences has had many other—essentially unpleasant and sometimes also deadly—consequences, both economic and personal.

The most obvious consequence is: Over a long period of time, entire societies were seriously negatively affected, practically all humanity. This was last the case with the so-called *Spanish flu*, which, by the way, did not come from Spain at all. Everywhere, people had to re-establish themselves within the context of the pandemic—families, schools, industry and other companies, artists, the health care system, craftsmen, retail, agriculture, tourism, aviation … simply everyone. Many of them reacted with improvisations and innovations (we will think about this in Chap. 3), some despair, too many moaned and complained, blamed the government, secret powers, doubted everything and everyone.

In the past decades, similar situations that were existentially threatening and psychologically difficult to cope with affected only a *limited* part of the world or society. In the

---

[3] https://www.reuters.com/article/health-coronavirus-italy-timing-idUSKBN27W1J2

[4] https://www.sciencealert.com/the-new-coronavirus-could-have-been-percolating-innocently-in-humans-for-years

[5] There are over 3000 different coronaviruses living on and in flying foxes alone (https://academic.oup.com/ve/article/3/1/vex012/3866407?login=false?), and WHO has indications (genetic traces) that racoon dogs may be the starting point from where the pandemic started: https://www.washingtonpost.com/science/2023/03/17/covid-origins-raccoon-dog/; there have been and there will be numerous occasions on which one or the other mutated virus could have jumped over long ago or will jump over to us humans in the future—what can this next virus then cause? Recent studies indicate to that racoon dogs might have spilled over from racoon dogs to humans, cf. https://www.scientificamerican.com/article/new-evidence-supports-animal-origin-of-covid-virus-through-raccoon-dogs/ (March 21, 2023), first publication in https://zenodo.org/record/7754299#.ZBvw1nbMLEY (March 20, 2023)

First and Second World Wars, Europe was directly affected, in the Second World War, large parts of Asia experienced the same. Indirect consequences were—and are increasingly now again due to the illegal and brutal Russian invasion of Ukraine—to be felt almost worldwide. But in the Covid-19 pandemic, no country was an island of bliss. When the coal crisis spread in the Ruhr region in the late 1950s and 1960s, this happened in Germany's largest conurbation. Not only were those directly employed in mining affected, but also the entire infrastructure, a large supplier and service industry. 20 years later, it had not been solved, but then it was overshadowed by the steel crisis, in between there was the oil price shock, the oil crisis.

The textile industry holds further examples of fateful coincidences that affect many people, but each only a part of the population. I don't want to go now into detail about the weaver's uprising—although of course, the development of the machines was anything but a planned and predictable process—no, this section of history was also characterized by accidental events, coincidences and enormous sudden breaks. In Germany alone, in recent times from the mid-1950s to 1980, almost half a million jobs were lost in the textile industry[6]—and this doesn't even include the affect on the suppliers and the entire trade, which had to adjust.

Then came, partially by chance, the *fall of the Berlin Wall,* which, of course, had many objective causes, but was triggered by apparently incorrect information announced by Günter Schabowski concerning a change in protocol for then-GDR citizens who wished to travel to the west. The reunification opened the way to a comprehensive social and economic renewal of the East German federal states, but, for many former GDR citizens, it was a deep depression: almost all seventeen million people had to orient themselves anew, and for many, this path is not yet finished or, in some cases, has not come to a good end. So, it is nothing new that large parts of society have to come to terms with new living and working/economic conditions and have to completely reorient themselves.

Why do I call the skipping of the virus to humans *coincidental,* and so adamantly? Well, first of all, we must be aware that a virus is not a living being, but needs organisms such as bats or humans to live—which, in the case of viruses, is limited to reproduction, because they have no metabolism. Viruses are in a gray area between inanimate and animate nature. A virus has no will, no goal—it lacks a brain and the corresponding nervous system to support it. It is only the *striving*—which is the wrong word, rather: its chemical reaction potential—of the RNA of the virus to double, and then to multiply by the millions, that drives it. In the sense of this *striving,* it would theoretically not be expedient to damage or even kill the host organism. It is purely coincidental, by accident, that it happens.

---

[6] https://de.wikipedia.org/wiki/Textilindustrie#Geschichte_in_Deutschland

Mankind, an Ecosystem for Viruses

There are billions of viruses on us and in us, about ten times as many as there are bacteria,[7] which, as single cells, are true living beings. According to everything we or the relevant scientists know, each one of us consists of about 100 trillion (i.e. 10^{14}) cells,[8] and, depending on the source, we also host 10 to 100 times more bacteria cells[9,10] or possibly "only" 1.3 times more.[11] That doesn't even include the ten times as many viruses, or the insanely high number of fungi that colonize us. We might well ask ourselves what we humans actually are. Maybe we're mainly an ecosystem that functions optimally by chance, one that happens to be perfectly suited for the real rulers of the world: microorganisms and viruses? But that's not a question for this book; we're thinking about chance and coincidence here. So, let me ask it another way: Maybe it's to be seen as an essential coincidence that we live, *despite* being inhabited by so many microorganisms and viruses? And maybe it's not by accident, but perfectly normal, that humans are threatened by influenza, Ebola, measles, and, since the end of 2019, SARS-CoV-2 viruses? However we want to look at it, accidental events and coincidence are definitely at work here, and not just a single one!

There are ten million viruses in every cubic centimeter of seawater, and it is estimated that there are a total of 10^{31} viruses in the world's oceans. These are estimated to be the representatives of about one hundred million virus types.[12] Viruses contributed almost ten percent of our genetic heritage during evolution.[13] As far as is known, the corresponding genes are mostly inactive. But we also know that a *retrovirus* that has been introduced into the genome releases proteins in only a few days old embryos, that is, shortly after fertilization, that prevent the entry of other viruses into the embryo, which, at this time, is still a cell mass. In addition, this former virus gene is involved in supporting the establishment of protein production in the cells of the embryo. Other former virus genes are active in placental cells and produce two proteins that "glue" the cells together.[14] Even parts of our genome that have been shown to come from bacteria probably came into our genome with the help of viruses.

---

[7] https://www.scientificamerican.com/article/viruses-can-help-us-as-well-as-harm-us/

[8] https://www.smithsonianmag.com/smart-news/there-are-372-trillion-cells-in-your-body-4941473/

[9] https://www.sciencefocus.com/the-human-body/human-microbiome/

[10] https://en.wikipedia.org/wiki/Microorganism

[11] https://www.scientificamerican.com/article/strange-but-true-humans-carry-more-bacterial-cells-than-human-ones/

[12] https://www.scientificamerican.com/article/viruses-can-help-us-as-well-as-harm-us/

[13] https://www.nature.com/articles/nature14308/

[14] https://www.spektrum.de/news/die-gute-seite-der-viren/1722318/

In our digestive system, not only can bacteria be found, but also countless viruses.[15] Their role is largely unexplored. Bacteriophages—a special family of viruses—are active in the intestine, mainly acting as our helpers by attacking pathogenic bacteria. However, viruses that can cause infections of the digestive system apparently also have a positive effect on the healing of damaged tissue in the gastrointestinal tract. Herpes viruses seem to be able to hinder bacterial infections. It is by no means clear what viruses, which are simply everywhere in the world, can do. But it is obvious that proteins enveloped in RNA—that is, viruses—have played a decisive role in the development of life. Indeed, it is not at all excluded that viruses or precursors of viruses are older than life itself, and thus were involved in the beginning of life.

In view of the incredible abundance of viruses of numerous types, of which the vast majority does not harm us humans and other living beings, but rather offers benefits or is neutral, it is certainly understandable what a truly unpredictable coincidence it was when the SARS-CoV-2 virus from the Corona virus family, together with its variants, became so virulent and dangerous for us. There are many experts and several previous studies that say that such a pandemic was to be expected. The close contact of people with wild animals, through hunting, wild animal markets, and our civil infrastructure penetrating deeper and deeper into formerly wild landscapes, is seen as conducive to this. Therefore, the probability increases that one of the trillion viruses surrounding us could become threatening again. Whether, when and how, and what kind of virus—that is all by accident. And to put it bluntly: This type of accidental event or coincidence, which happened at the beginning of the pandemic, is an essential characteristic and tool of evolution. This concatenation of coincidences, at the beginning of each of which are mutations, over millennia and millions of years, we call *evolution*.

Coincidental Inventions and Disasters

This is also the case with events that change technology, science or history. Actually, most things are set in motion by chance, by accident. This is the case with all inventions. But what exactly is an invention? It is a new idea that is then implemented worldwide for the first time by means of work—often very systematic, very time-consuming, rarely very quick and easy—in industrial practice, in the market, in our households. Ideas form purely by chance (I will come back to this in Chap. 3). Anyway, something happens in the brain while inventing: Suddenly, you see how a problem can be solved in a new, previously unthinkable, often even inconceivable way. Nobody can plan or even force an idea, no, it's not that simple. Such ideas come to you, or they don't.

I have made several dozen inventions of different kinds myself. And I've had even more ideas. The ideas all came to me unannounced and mostly in situations where I

---

[15] https://www.frontiersin.org/articles/10.3389/fmicb.2015.00918/full

was thinking about anything but the problem to be solved: while doing crane research in nature, during breaks in my soccer games, while cycling, or on the toilet. The invention ideas came to me in the form of images, not in the form of words or sentences. Of course, my inventions, of which a large part were also patented, and some were successfully introduced into the market, were not so world-changing that you would necessarily have heard or read about them before (although you are most probably using them, without knowing, you will see it later in the book). But you know other inventions, and they all had happened by chance.

This is how the 3M corporation is known to most people: because of the small yellow notes that can be stuck to a wall, a glass, a piece of paper, a free space on your desk, or on the shade of your desk lamp. Most of the time, these notes are about small things like the one phone call you need to make at 4 p.m., or that you should buy vegetables on the way home. When the job is done, you take the note off again; it's very easy and leaves practically no residue on the surface, quite different from if you had done the same with conventional adhesives. Brilliant, isn't it? But how was the product, for which the patents have long expired, created? Well, the relevant engineer wanted to invent a superglue. He didn't succeed, he only managed to create a very weak adhesive, which 3M wanted to market as a pin-wall without pins, which ended up in a failure. This adhesive, which didn't stick, was forgotten for six years. Until, suddenly, a colleague of the "inventor" who sang in a church choir got annoyed that his bookmarks and notes kept falling out. That's when he got the idea to dig out this forgotten adhesive. So, the ubiquitous "Post-it"[16] was born.

Melitta Bentz, whose husband operated a household goods store, was annoyed by the coffee grounds in her coffee or, alternatively, by the constantly clogging linen towels that were placed in normal sieves. She pierced a tin can bottom and placed a circle-shaped piece of blotting paper inside. The first Melitta filter had been created. Later, the filters and filter papers were developed as we know them today. Based on this idea, an entire big company grew.

Viagra was the result of a failure of research aimed at designing a drug to deal with high blood pressure. It was only because several test subjects, after the test was supposed to be stopped, repeatedly asked whether they could not continue receiving replacement after all, that Pfizer accordingly investigated what on earth, if there had been no positive developments related to blood pressure at all, was so interesting to the male test subjects (while not to females, no woman had also asked for further supply of the experimental pills), a potency drug and a billion-dollar business emerged from the failure.

On April 11, 1970, the third moon landing expedition, Apollo 13, took off. If you look at the complete history of this flight, which was plagued by bad luck,[17] you may be surprised that at least some things went according to plan. Almost everything that happened

---

[16] registered trademark of the company 3M.

[17] https://www.nasa.gov/mission_pages/apollo/missions/apollo13.html

was unplanned. It was not only that one of the crew members suddenly developed measles and a replacement had to step in (the latter, of course, was happy because he was able to join the flight). There is not enough space here to list all the things that went wrong before and during the launch. But then, after almost 56 hours of flight, an oxygen tank exploded. The Apollo capsule lost air. The crew escaped to the lunar module, which was originally only intended for two men (for the purpose of landing on and returning from the moon's surface). The problem (besides the lack of space) was not the oxygen supply, but the removal of the exhaled carbon dioxide. The filter in the lunar module was not designed for three people and not for such a long flight. With some improvised tinkering, it was possible to sufficiently expand the CO_2 filtration capacity. Improvisation, the solution of problems using limited and, often, insufficient means, requires creativity, a whole lot of ideas: I will come back to the role of coincidence in the emergence of ideas later.

Even catastrophes are ultimately coincidences or the result of a series of unfortunate coincidences. On March 11, 2011, an earthquake occurred 163 km east of the Fukushima nuclear power plant in the sea off Japan.[18] There, a piece of the Pacific plate had subsided, causing a magnitude 9 quake, the strongest earthquake that had ever been recorded in Japan.[19] Thus, it happened a) suddenly and unexpectedly and b) in a magnitude never observed before. The seismometers triggered a fast shutdown of blocks 1 to 3. The external power supply failed, except for one unit, and all emergency power supplies kicked in. All six blocks switched to emergency cooling. The earthquake shook the power plant blocks much more than had been assumed possible in the planning of the plants. About fifty minutes after the earthquake, tsunami waves with a height of 13 to 15 m hit the coast; 19,000 people lost their lives as a result of the force and the devastation. The waves also hit the power plant. The protective wall was only slightly more than five meters high,[20] and only slightly more than three meters had been prescribed. The sea water pumps were destroyed, waste heat could not be dissipated, and five of the twelve emergency power supplies and the distribution cabinets were flooded. Simultaneous accidents in several power plant blocks were not foreseen in the emergency plan. There was overheating, release of hydrogen and core meltdown in blocks 1 to 3. Hydrogen explosions followed in the next days and destroyed the buildings of four blocks. In the subsequent days, additional problems were caused by further accidents. For example, a rat gnawed on a cable insulation, which led to a short circuit that shut down the provisionally repaired cooling systems for the cooling ponds.[21] It was not until six years after the core meltdown that the first melted fuel rods in the form of solid *lava* were

---

[18] https://en.wikipedia.org/wiki/Fukushima_nuclear_disaster

[19] http://bit.ly/366vkpE (focus.de).

[20] another source (https://atomkraftwerkeplag.wikia.org/de/wiki/Der_GAU_von_Fukushima_und_die_Folgen) speaks of 10 m, which was also not sufficient.

[21] http://bit.ly/366vkpE (focus.de).

found.[22] 160,000 people had to be evacuated. The soil is contaminated for decades or centuries to come. Two thousand workers were contaminated. These are just some of the consequences. My main point with this example is to show how even experts fare poorly when it comes to calculating and adequately taking into account the effects of coincidental natural disasters. The earthquake happened to be extremely strong, the strongest ever observed in this region. The tsunami was higher than ever thought possible. The emergency measures were not up to the forces of nature. The robots used to examine the extremely contaminated reactor buildings with the melted fuel elements also succumbed to the strong radiation before they could deliver results. Only the third generation of robots—developed ten years after the disaster—lasted long enough in those conditions to make further explorations in the shallows of the reactors possible.

This is how it is with natural disasters. After they have happened, we are always smarter, it is commonly said. But is that really the case? Because the next disaster behaves just as differently and is just as unforeseeable. Of course, there are damaging events that are relatively easy to predict, if, at the very least, we have previously experienced similar events. If we clear and develop the mountain slopes, seal our country's soil at an enormous speed, straighten the streams and rivers, seal off with dikes the meadows and forests that were previously flooded, it should come as no wonder and no coincidence that, after heavy rain, floods occur somewhere at the lower reaches of the river. And if the houses are built right up to the banks, the cellars and living rooms fill up, or they are completely destroyed, as happened in 2021 in the German Ahr Valley. This in a valley that has known flood catastrophes for centuries, but where houses have nevertheless been built further and further up to the banks. If we quickly drain huge amounts of the surface water resulting from rain via ditches, canals and rivers into the North Sea, and if we simultaneously extract more groundwater than we allow to recharge, we should not be surprised that the groundwater level is falling. So, it is by no means the case that we cannot calculate anything or at least roughly predict it. What we cannot predict, cannot calculate, is the kind of accidents and coincidence that is based on complex, interlocking causes that we usually only understand afterwards at best.

Chance in Economy and Politics and the Unpredictable Consequences

It is not only in regard to natural phenomena that we find such accidental events. We should all know this by now, decades of experience should have demonstrated it to us: The stock market, the development of financial markets, our German, European and even world economies are basically not predictable. With all these constantly published forecasts, it is actually the case that approximately half of the forecasters say A, the other

---

[22] https://www.nytimes.com/2017/11/19/science/japan-fukushima-nuclear-meltdown-fuel.html

half B. And, of course, with the huge range of predictions, one of the *experts* has hit the jackpot and can be celebrated as a real soothsayer. But all of this is not even coffee-table guesswork. Nevertheless, there are strongly competing organizations obsessed with reading the coffee grounds. Even without particularly exciting unpredictable global or specifically European or German events, the stock index or the currency rate can stagnate, fall or rise compared to other heavyweights. The causes are often not transparent—simply because there are too many actors who trade too many securities, making the whole thing unmanagable and anything but predictable. A company, or an entire country can get into trouble if a weighty crowd of trading participants comes to the opinion that a sloping position is darkly threatening or huge opportunities at this or that company or industry are promisingly shining on the horizon.

These financial and economic figures are particularly influenced by significant events. The effects of the Corona pandemic, which was not foreseen, began with an unpredictable coincidence of zoonosis and has slowed down—as I write these lines. In Europe, the Brexit will also contribute to this, a political earthquake that no one foresaw: When the referendum was scheduled, then-UK Prime Minister David Cameron was absolutely sure that Great Britain would vote to remain. Even the Brexit supporters, led by Boris Johnson, did not believe in majority support for leaving the EU. Probably because they simply did not believe in their own lies. When the voters nevertheless voted for Brexit with a razor-thin majority, the pro-Brexit politicians were caught off guard and were not prepared for the negotiation with the EU as to how and when the exit should take place. There was no big follow-up coincidence after the accidental result in the referendum, but it will remain a source of further surprises and coincidences. The consequences of Russia's attack on Ukraine and the ensuing sanctions were completely unpredictable at the time when I was adding these lines a few days after the war began, and the outcome is still completely open now in 2023. Apart from the US intelligence service, all politicians and observers were probably convinced that a war could be avoided, and we ordinary citizens hoped so. A misjudgment. Later, we will read a lot about how to understand what happened politically and economically, as well as in the financial world, afterwards—but there are no secure predictions now. There are plenty of assumptions, and predictions too, but we should not be surprised if, after a few years, we are all surprised, the experts just as much as anyone else, that everything will have developed differently than expected, differently than predicted.

We see things more clearly *after* the unforeseen event of September 11, 2001, when terrorists steered two planes into the towers of the World Trade Center. Osama bin Laden had not anticipated that the towers would collapse completely, but only that the floors *above* the point of impact of the planes would catch on fire.[23] Consequently, it was not foreseeable that there would be thousands of deaths due to the complete collapse, which

---

[23] https://edition.cnn.com/2002/US/09/10/ar911.osama.exclusive/, in this article, it says: "Due to my experience in this field, I was thinking that the fire from the gas in the plane would melt the iron structure of the building and collapse the area where the plane hit and all the floors above it only. This is all we had hoped for," bin Laden said on the tape.

lasted only a few seconds. But, of course, the stock market dropped drastically, currency rates went crazy, the economy suffered severe losses. The Swiss franc, on the other hand, strengthened—but why? Are the irrational assumptions of many currency market participants that the Swiss franc is a *safe haven* in such catastrophic events just a coincidence? What could this *haven* possibly offer? There is nothing tangible in sight. This all happened in a matter of minutes. After a few months, the rates and the economy had more or less normalized again. But over the following two years, my company *Ormecon* just barely avoided bankruptcy several times. However, the global political consequences can still be felt today, especially in Iraq and Afghanistan and with the new terrorist organization *Islamic State.*

Ultimately, however, September 11 was not the starting point of this development, but rather a later step in a longer, complex process that began after a very short-sighted decision by the British in 1951. Here, we see how "a tornado can be set off by the flap of a butterfly's wing" (we will return to this metaphor in Chap. 6):

Iran had a secular and left-wing government that, in 1951, decided to nationalize the oil industry.[24] The British, who owned a majority in Iran's largest oil company, imposed a naval blockade. They persuaded the US government to foment a coup against the democratically-elected government and install the king of this parliamentary monarchy as an absolute ruler, as *Shah of Persia.* To do this, they claimed that the Iranian government was leaning towards the Soviets, a situation that had to be nipped in the bud with all due speed. From the perspective of the USA and the whole West, this coup looked advantageous for a long time. But below the Western perception threshold, dissatisfaction with the Shah's luxury and his Western orientation increased steadily, especially in view of the contrast to the low living standard of most Iranians. In addition, there was the brutal behavior of the secret police. This eventually led to Ayatollah Khomeini gaining influence—the rest we all know. The revolution initiated by him later inspired other Islamic movements and Islamist terrorism. President Barack Obama later classified the original wrong decision—to overthrow the democratically-elected government and install the Shah—as one of a series of similar wrong decisions that I would call *coincidences in history.* Coincidences because there were several—from the perspective of the probability of occurrence—at least equally good decision options. One of them is the most obvious: non-interference in the internal affairs of other countries.

In the case of the developments in Iran, which were then ultimately one of the main causes of Jihadism, for example, one could, of course, also argue: This decision of the US and British governments to organize a coup in a foreign country was based simply upon the then-political orientation and interests. That's right—but was it therefore inevi-

---

[24] The following information is from Barack Obama, A Promised Land, Penguin 2020, pp. 629 ff. in the German edition "Ein verheißenes Land", Penguin-Verlag 2020), and, of course, it was all much more complicated than represented by Obama and certainly much more than summarized here by me, with a whole lot of other coincidences playing a role.

table? It certainly wasn't, because the interference of the USA and Great Britain also violated international law at the time, and the operation was kept secret for a long time. We can be sure that there were dissenting voices in both governments. So, there were enough alternative options for both Western governments, not least to do nothing at all. The decision for the coup finally comes about by chance, namely, based on how many influential people decide in favor of this option. As well as what the reasons were for or against it for each individual. And then it's just a matter of chance whether the majority of those entitled to vote are in favor of an action or against it.

Most events that influence the course of the economy are similar and can be classified. In retrospect, it often looks as if it could not have happened any other way than it did. And after an event has taken place, the majority of comments in the media sound as if everyone knew in advance that it would happen, no: it *had* to happen. Only, interestingly, nobody has read or heard about it anywhere before. "In retrospect, everything makes sense," writes Daniel Kahneman in the already quoted book *Thinking, Fast and Slow,*[25] continuing "And we cannot suppress the powerful intuition that what makes sense in retrospect today was predictable yesterday. The illusion that we understand the past encourages us to overestimate our ability to predict the future." Peter Ustinov once came up with a delicious bon mot: "It's best to talk to prophets three years later."[26]

Of course, there have always been experts who predicted what actually happened, but the majority of experts are usually completely wrong. This was also the case with the financial crisis that culminated at the end of 2008: In mid-September, the large US bank Lehman Brothers collapsed, as a result of which the entire global financial system could have collapsed if the central banks had not intervened massively. Even today, the consequences can be felt in the global economy and in the financial systems.

For me personally and my company with our new product—especially in the Chinese printed circuit board market—this crisis was also of great importance. Of course, I knew that the economic situation was uncertain, but somehow we managed to muddle through. After the setback in 2001, I had brought the company back on track together with my employees, a feat that required, in addition to the courage to take risks and hard work, several lucky coincidences. These had led first individual customers and then me, in 2005, to China. Since spring 2008, I had been trying to sell the company, and was negotiating with a medium-sized US stock corporation that had a strong Asian subsidiary and did a lot of business in China. At the end of August, we were moving towards an agreement, the contract being essentially in place—there were just some nitpicky concerns on the part of the Americans, and my fellow co-shareholders had even more concerns that were even less important. They said we should rather wait a little longer and "think everything through carefully and well". But we two CEOs wanted to conclude *immediately* and not after "all concerns had been resolved", which would have taken forever. We had

---

[25] Daniel Kahneman, "Schnelles Denken, langsames Denken", loc. cit. (German edition), p. 270.
[26] gutezitate.com/zitat/225135.

dispelled so many points of contention, found so many compromises —now it had to be over. We agreed to threaten our boards: "If we don't sign very quickly now, the other side will shut down." There was absolutely no talk of this on the other side, but we took this risk of the alleged ultimatum on the other side. It worked as hoped. The groundless concerns were dropped.

At the beginning of September, we met with a Hamburg notary public for the signing. Exactly ten days later, the US bank Lehman Brothers collapsed. My negotiating partner and I met again in China the day after the bank disaster, and he candidly admitted: "If we hadn't signed a few days ago, the deal would never have come about. Corporate management is now in sheer panic mode." We had simply been lucky that we could sweep the pusillanimous concerns of the approval committees off the table. Sai Weng was right with his philosophy of life.

2009 was an economically terrible year. But already towards the end of this catastrophic year, and even more so later, it became clear that the contract with the US company was a stroke of luck, because, within a very short time, we developed my process into the company's most sales- and profit-strong product. Only a few years later, we were market leader with my product. *We* then meant, however, the buyer of my company, while I was working in a completely independent position (not subject to any instructions) as a technology consultant for the group, although I was actually leading all the technical marketing and laboratory work for my product, together with my former employees in Germany and China. Above all, I advised the customers and their customers, the international large corporations of the electronics industry, in terms of application technology. If we had delayed the signing of the contract by a mere ten days, the sale would have collapsed. In the following years, the US group would probably have encountered enormous problems without our product, because all other product groups shrank or, at best, stagnated in the course of the next years. And whether I would have survived the years 2009/2010 in China with my small company is definitely not written in the stars.

The *Financial Crisis Inquiry Commission* legally installed in 2009 took two years to uncover the causes of the financial crisis. It turned out that there were numerous, long-term developing and mutually interacting causes that eventually brought about the collapse. It is the classic toolbox for the generation of coincidences: The individual reasons would never have caused such a world financial crisis on their own. And because no one was able to foresee that this and that and further reasons would cause such a fundamental crisis of the entire world financial system several years later, we also see here the very typical composition of the type of coincidence that we want to examine in this book.

Accidental Events and Coincidences in Scientific Research

It is no different in natural scientific research, even if researchers themselves like to tell their story differently in cases of success—namely, that everything actually followed a great long-term plan that was systematically implemented by oneself or the department

or company. Of course there had been failures, that's normal, but after patient systematic work, success finally set in. You may believe this or not; I don't believe it: The vast
majority of research results have been highly dependent on chance and coincidence. On
the fact that the researchers met people important to their work, that they were lucky
enough to find good employees, that they were able to prevail with a research application, not least because they *accidentally* heard or read or observed something that helped
them decisively. Only very rarely does one read anything from researchers themselves
about how much their research results depended on chance and coincidental events.

James Watson, on the other hand, very illustratively described in his book *The Double
Helix*,[27] how convoluted the path to the discovery of the DNA structure was. I devoured
the book as an 18-year-old aspiring student who wanted to become a biochemist. It still
sits on one of my bookshelves today. Watson and Crick were already my idols during my
high school years. I know that, in recent years, Watson has maneuvered himself onto the
sidelines with various controversial (racist and homophobic) remarks. Researchers who
would express similar views today would certainly not be given a leg to stand on. But
Watson's later unquestionably rejectable and reactionary views do not change the significance of his and Crick's pioneering achievement.

His book is extremely enjoyable to read. Watson was born in the USA, grew up
there and went to university at the age of fifteen. He is highly intelligent. At first, he
wanted to specialize in ornithology, but then switched to zoology because of his interest in genetics. He began his first research project in a group at Indiana University that
was determined to clarify the chemical basis of genes. Many coincidences eventually
led him from Copenhagen (where he was a postdoc and was actually supposed to be
working on another topic) to Cambridge, bringing him into contact with Francis Crick.
Thus, two congenial scientists, who were also unconventional in every respect and liked
to stray from the usual paths, met by chance. They did their mostly mental and modelling research work not with the blessing of their bosses, but outside of their actual tasks.
Not least, they did not like to deal with chemistry at all, although they had set themselves a structural-chemical question. At that time, there was one hypothesis, or let's call
it a *paradigm* (more on this in Chap. 5), that enjoyed the greatest popularity, according to which the genes in the chromosomes consist of proteins. Deoxyribonucleic acid
(DNA) was considered an unimportant interfering by-product. But there were also first
signs and individual researchers who suspected that genetic information was housed in
the DNA. Watson's and Crick's chaotic, playful and creative approach is clearly visible in Watson's book. And one of their main motives—in addition to the acquisition of
knowledge—was to be faster in the elucidation of the DNA structure than the already
famous Linus Pauling (he received the Nobel Prize in 1954). Pauling had recognised the

---

[27] J. D. Watson, The Double Helix, Atheneum Press (US) and Weidenfeld & Nicolson (UK) 1968.

helix ("alpha helix") as the secondary structure of proteins.[28] This prompted Watson to assume a secondary structure for DNA as well. Here, in the sense of our book, we see a key coincidence, the unproven flash of inspiration that is not proven by anything: *Maybe DNA also has the form of a helix, like proteins.* But two scientific prejudices still had to be overcome. On the one hand, that they really focussed on DNA, although, according to common research opinion, proteins were considered carriers of genetic information. On the other hand, what a DNA helix could look like; because they assumed that the phosphate groups were on the inside, the bases on the outside. When they learned that Pauling was going to publish a new model of DNA, Watson and Crick finally got the approval to focus their research mainly on elucidating the DNA structure. Only then did Watson turn his firmly anchored model idea around and placed the bases in the interior of a double helix in his structural model, and the phosphate groups to the outside.

In Watson's book, it becomes very clear how, over the course of time, accidental hints from other scientists created small forks in the road to knowledge, which corrected false assumptions about details. Not least of all, fortunate coincidence played a role in Maurice Wilkins showing them X-ray structure spectra of Rosalind Franklin. Watson and Crick used the data from Franklin and Wilkins for their double helix model without asking. Watson, Crick and Wilkins received the Nobel Prize in 1962, while Rosalind Franklin went away empty-handed, having already died of cancer in 1958.

逆水行舟, 不进则退 (Nì shuǐ xíng zhōu, bù jìn zé tuì, "Boat [fights] against the current, [if] not forward, then [one falls] back") is a Chinese proverb. It goes back to the ancient Chinese philosopher XunZi, who lived more than two hundred years before our time. It is generally interpreted as follows: Only those who swim against the current will eventually reach the source. I understand it this way: You can only really understand the origin of things if you fight your way to the roots against all resistance. Or the other way around: You can't achieve anything new if you just follow the mainstream. Watson and Crick swam against the current.

Wilhelm Conrad Röntgen, not very successful with his research during his many years as a professor, was playing around with a new invention, a *cathode ray tube*. It was still in its infancy then, and what one could do with it, if anything at all, was completely unclear. Röntgen wanted to find out why these tubes glowed so strangely. It was not known that electrons were accelerated in these tubes, because electrons had not yet been discovered.[29] Strangely enough, a piece of paper on which fluorescent paint was applied began to glow when he turned on the tube. Röntgen was annoyed, and wrapped the tube

---

[28] By "secondary structure" (e.g. in proteins) we mean that the chain of amino acids ("primary structure") forms in stretches into helical sections, after which the entire chain then folds spatially and forms a three-dimensional superstructure ("tertiary structure").

[29] https://www.scientificamerican.com/article/the-discovery-of-x-rays/

in black cardboard. The paper still lit up. He had discovered an invisible radiation. Then, he experimented on and on until the next discovery came: These mysterious *x-rays,* still called *x-rays* in English today, left traces on light-sensitive photographic paper that could be developed. With the photo of his wife's x-rayed hand, in which the bones were visible, the breakthrough also came in the popular media. Once again, it was a series of coincidences that eventually brought us the X-ray machines and later, for scientists, the X-ray structural analysis (from which Franklin, and thus Watson, also benefited).

Many groundbreaking discoveries in research have been made by chance or after a sudden inspiration. The famous mathematician and physicist Roger Penrose, while crossing the road, came up with the idea, right in the middle of the street, as to how he could prove with a calculation that black holes exist, and why.[30] He published this proof when it was still very controversial as to whether such terrifying giants actually existed. And his argument was viewed as a theoretical gimmick for the critics for so long—until black holes were actually proven to exist. In 2020, Penrose was awarded the Nobel Prize, together with Andrea Ghez and Reinhard Genzel. The latter provided evidence that a supermassive black hole, a voracious monster, is located in the center of the Milky Way. A brainstorm like the one Penrose had is also a coincidence, an event happening by chance in the brain. We will look at this again in a little more detail later.

Such a coincidence, a brain flash à la Penrose in the middle of the street, is not just pure luck. The brain must be prepared, it must—as in the example with Penrose—be mathematically trained, it must have thought intensively about the problem beforehand. Then, the *inspiration flash* can strike and become conscious. There is a term in American English that describes such coincidences in research and invention better than the term *lucky chance: serendipity.*[31] The term means that only the "prepared mind" is receptive to unexpected discoveries, as Louis Pasteur once expressed it: "In the field of observation, lucky chance favors only the prepared mind."[32] Serendipity is the art of recognizing something unexpected as relevant that one encounters by chance while searching for other things.

Let's linger a little longer on black holes. In his gripping book *Light in the Darkness*, Heino Falcke describes how he managed to bring together a very large, global team and all the financing and infrastructure required to finally take the first *photo* of a black hole in the center of the distant galaxy M87.[33] It is six and a half billion times heavier than

---

[30] https://www.nobelprize.org/prizes/physics/2020/popular-information/

[31] https://www.scientificamerican.com/article/the-beatles-may-owe-some-of-their-success-to-pure-serendipity/

[32] original: "Dans les champs de l'observation, le hasard ne favorise que les esprits préparés." https://www.curieuseshistoires.net/louis-pasteur-medecin-vaccins/

[33] Heino Falcke (with Jörg Römer), "Light in the Darkness: Black Holes, the Universe, and Us", HarperOne 2021, original German edition "Licht im Dunkel" by Klett-Cotta 2020, citations taken from the German edition.

our sun. Prof. Falcke's chain of fortunate chances in the sense of serendipity begins with his leafing through an old conference volume and coming across a publication. After reading it, he develops the idea of how to make *the shadow* of a black hole visible. A chance encounter on the train on the way back from a conference finally leads to successful research funding application, which, thanks to a risky sub-project, was awarded the contract:[34] The search for pulsars in the center of the Milky Way had been included in the application. For decades, the world had been searching for them, a search that had had so far been unsuccessful, although thousands of such special neutron stars were suspected to be there. During the application review period, *suddenly* ("what a coincidence!"), a new pulsar began broadcasting radio pulses. The team at the Max Planck Institute for Radio Astronomy in Bonn, to which Falcke also belonged, discovered and measured this pulsar as the first in the world, published the sensation in *Nature* and thus received international attention.[35] Needless to say, this grandiose coincidence provided a timely vote of confidence for the research team applying for the grant.

Before their second final rehearsal, the scientists learned of another premiere: the first measurement of gravitational waves. For the first time, at the end of a test run, the LIGO experiment was able to collect data that, after evaluation, turned out to be the trembling of space predicted by Einstein. Two black holes had emitted these signals during their merger. What a lucky coincidence: If the facility, at that time only doing a test run, had been shut down a few hours earlier, they would not have been able to pick up the signal. Falcke wrote about it at the time: "We'll never be that lucky!"[36] But they were, not only on the days of the worldwide measurements, but already during the many years before.

Before the world premiere to make a *photo*[37] of a black hole was successfully brought about, a whole series of other fortunate coincidences was required. The most important stroke of luck was provided by the weather during the decisive measurement campaign, because the radio telescopes located all over the world and precisely interconnected via atomic clocks all needed to experience good weather at the same time, and that for days! During another planned measurement campaign, they had bad luck—the weather did not play along. But the first attempt in 2017 was enough: The black hole *Sagittarius A** could be detected by means of its shadow. The image published in 2019 was then the result of two years of evaluation (in part, on supercomputers) of three thousand terabytes of data on one thousand hard drives, which the eight interconnected radio telescopes had collected on four continents, including at the South Pole, during the five-day measurement period.

---

[34] H. Falcke, loc. cit. (German edition), p. 198 to 206.

[35] https://www.nature.com/articles/nature12499

[36] H. Falcke loc. cit. (German edition), p. 216.

[37] which of course is not a photo in the usual sense, but a visualization of radio waves.

The Essential Coincidence—an Attempt at Description

"The physical research of the last 4–5 decades has shown, clear as daylight, that, at the very least, for the overwhelming majority of events whose regularity and constancy have led to the postulate of general causality, the common root of the observed strict regularity—is chance."

Erwin Schrödinger: *What is a natural law?*[38]

This sentence comes from Erwin Schrödinger's inaugural speech at the University of Zurich in 1922. Even if Schrödinger probably meant other types of chances and coincidences than the ones we are examining in this book, the sentence, if we take it literally, is completely correct. In the hundred years since he spoke it, we have only learned much more about chance and coincidence.

We should now try, after reading several examples and Schrödinger's statement, to find a general definition for the type of coincidence that I want to limit myself to in this book.[39] There are several types of chances/coincidences. For example, there are those for which we—who experience them—do not recognize a direct cause ourselves, but only because we have overlooked them. So, we do not have to complain about the alleged coincidence that we have not changed our bicycle tire for years, now that it has gotten a flat tire on a gravel road. It may be a coincidence that it happened *today*, but it was basically foreseeable.

The eminent biochemist and Nobel laureate Jacques Monod examines the concept of chance and coincidence somewhat systematically in his book *Chance and Necessity.*[40]

---

[38] From: "The Natural Sciences" Issue 1, 04.01.1929, p. 9–11; the pdf of this speech can be downloaded here: http://www.psiquadrat.de/downloads/schroedinger22_29_naturgesetz.pdf. The phrase was translated from German by B. Wessling.

[39] In this book, we will **not** deal with the question of how people in general perceive chance and coincidence or accidental events, or the fact that that we humans also like and (pre)tend to see connections and supposed causes where there are no connections and no objective causes (belief in supernatural powers and forces, superstition, esoteric beliefs, conspiracy theories).

[40] Jacques Monod, "Chance and Necessity—An Essay on the Natural Philosophy of Modern Biology"; (German edition "Zufall und Notwendigkeit": Piper-Verlag p. 107, page numbers corresponding to the edition of 1996; I had already bought and read the original edition from 1971 at the beginning of the 1980s, but after 10 more moves, I could not find it again in my bookshelves, so I bought the antiquarian edition of 1996 for the work on this book). The original wording in French is "hasard essentiel", cf. Jacques Monod, "Le Hasard et La Nécessité—essai sur la philosophie naturelle de la biologie moderne", Éditions du Seuil, Paris, 1970, p. 128, here also the term "incertitude essentielle" is used and explained. In the following English edition ("Chance & Necessity"—An Essay on the Natural Philosophy of Modern Biology", Vintage Books, 1972), the terms "essential uncertainty" and "essential chance" is used and explained, p. 113 and 115, resp. Instead of "chance", I mainly prefer to use "coincidence", in addition to "chance" and "accidental events".

He starts from a quote by Democritus: "Everything that exists in the world is the fruit of chance and necessity."[41] But then he differentiates between chances in contexts that can be investigated using probability theory and those phenomena to which chance or coincidence is "essentially" inherent.

The first case we encounter, for example, is with roulette, the unpredictability of which, according to Monod, has a purely operational, technical cause. It is rather impractical for us to predict where the ball will land. There have been attempts to outsmart the ball's random behaviour in the roulette wheel. J. Doyne Farmer developed an appropriate procedure in the 1970s.[42] A computer with suitable software must be told a) how quickly the ball can traverse the circumference of the wheel one time, and b) how long the wheel takes for one rotation. The inventor secretly measured this with a stopwatch and transmitted it by radio to a partner outside the casino. This allowed the computer to calculate quickly before "rien ne va plus". With this approach, the chance of winning by placing a bet on, for example, "red" increased from 48.6% (which means that you lose in the long run) to 51.6% (so that you win in the long run if there are enough attempts). In theory, of course, an (undesirable) apparatus is also conceivable, one that uses laser beams to carry out the measurements much more accurately and quickly. The calculation would be carried out today with a modern fast laptop instead of a slow 1970s computer, a change that would increase the chances of winning even more. Of course, this could never be practiced, but J. Doyne Farmer showed that roulette does *not* generate *essential chance*. What we observe with roulette, is *randomness*.

The situation is similar with dice or other games, and, in principle, also with the lottery: randomness. With *6 out of 49,* it is *not* fundamentally impossible to predict which ball will fall next by means of a super-exact observation of the flight paths of all the balls. Only, it would not help anyone to win, because you have to submit your lottery ticket the day before. But the probability of winning is exactly known.

It is just as apparent that my followers on Instagram have, on average, more subscribers than I do; the same is true of most of the photographers I am connected to: If they checked, they would observe that their contacts, on average, have more contacts than they do. In the case of Facebook, it was calculated in 2011: At the time, Facebook users had, on average, 190 friends; but the friends of this average user had, on average, 635 friends. For 93% of the people on Facebook, their own list of friends is shorter than the list of friends of their friends. This appears illogical, but it is so. Scott Feld has mathematically examined it using statistics and analysis of the "sample bias" principle in his paper *Why your friends have more friends than you.*[43]

---

[41] J. Monod, loc. cit (German edition), p. 17.

[42] https://www.sciencealert.com/a-physicist-has-built-a-machine-that-can-beat-the-odds-at-roulette

[43] Scott L. Feld, "Why Your Friends Have More Friends Than You Do", American Journal of Sociology, Volume 96, Number 6, May, 1991, pp. 1464–1477. Numbers of the Facebook study cited after Christian Hesse, Warum deine Freunde mehr Freunde haben als du, Springer Nature 2017

It is quite different with coincidence as it occurs in evolution; for this, coincidence is an inherent or *essential* property, according to Monod. To explain the principle of *essential coincidence*[44] (Monod is using the term "hasard essentiel" in his original French book, and "essential chance" in the corresponding English edition[40]), he first draws a constructed example: A doctor is called to a patient; a plumber is busy with a repair on the roof of the neighboring house; the plumber carelessly drops the hammer, which falls on the head of the doctor, who dies on the spot (one could spin Monod's story further and say that the patient also dies because he did not receive the urgently needed insulin injection in time).

The coincidence is therefore *essential* because the two event series are completely independent of each other. On the one hand, the doctor is on his way because he was called, and he is perhaps only below the point where the hammer will fall in that exact second because he turned back again, for example, to retrieve his forgotten umbrella; on the other hand, the carelessness or even fatigue of the plumber has absolutely nothing to do with the doctor's path. Such coincidences cannot be predicted with a certain probability, unlike the winning chance in roulette, a dice game or a lottery.

It is similar with the sequence of coincidences that brought me to the job with Zipperling in Ahrensburg: The event chain that led me to print an advertisement in the FAZ Wednesday edition had nothing, absolutely nothing to do with the event chain that led to Mr. Petersen being bored and unable to fall asleep that Wednesday night, not to mention reading the academic job advertisements for the first and only time in his life. If one wanted to calculate the probability of this chance, a value of practically 0% would result. This is then an essential coincidence.

And so it is with evolution: The chance that there is a copying error when replicating DNA has nothing to do with the subsequent coincidental events. The genetic information is first translated into the synthesis of enzymes; these are proteins that then form very complex three-dimensional structures. What this folding will look like, what the resulting enzymes can do better or differently, whether they lead to fatal malformations, and, above all, what consequences this, in turn, has over countless similar accidental steps leading to the emergence of new species, is not predictable, not calculable. A probability consideration would also result in infinitesimally low values close to zero here. But without such coincidences, there would be no evolution, they are essential.

In other words: The result of a mutation, that is, a change in the composition of the bases[45] in DNA, usually at only one single point, is completely unpredictable.

---

[44] Just as an "essential amino acid" is so called because we cannot exist without it (our body cannot produce it itself), "essential chance" is to be understood as meaning that processes that have such types of chance would not function without them.

[45] Just as a reminder from biology class: The chain and helix structure of DNA is formed by certain sugars to which phosphate groups are attached. Inside the double helix are organic bases, so-called purine and pyrimidine bases. Three bases in a row (a "base triplet") encodes a specific amino acid; for a specific amino acid, there are between two and four different triplet arrangements, see also https://en.wikipedia.org/wiki/Genetic_code.

For example, one base could be "missed" when copying; then, the triplet is read completely differently, a *completely* new protein is created. This might be simply useless or it might be harmful, or it could cause a new property of the organism. On the other hand, if only one base is read incorrectly, this might have no effect at all, or it could lead to a completely different folding of the protein. What this, in turn, does to the organism is also unpredictable. Apart from that, a new species is not created by a single mutation, but only over the course of many generations, with many equally accidental mutations and the coincidentally resulting changes.

So, we are dealing with a chain of coincidences in evolution that are initially unrelated to each other: The accidental—and extremely rare!—occurrence of a copying or reading error; the resulting coincidental change in the now differently composed protein; the then also resulting unpredictable change of a property, unless it leads to a disease or the death of the organism, which, in turn, influences the ability of the organism to survive and reproduce in the ever-changing environment.

It is this type of intrinsic or *essential* coincidences based on "essential uncertainty" (as Monod worded it) that I want to examine in more detail in this book. We will not limit ourselves to mutations and evolution. Therefore, we want to rethink the term *coincidence, or chance*, again. Of course, it is worth taking a closer look at philosophical definitions of coincidence (chance) at the beginning.[46, 47] However, I don't want to adopt such formulations, but rather describe it with my own words as follows. That is, I *do not define* it, because there are certainly people who would be more competent at this sort of lexical definition, but formulate it for our purposes in this book so that we can understand it better:

Essential coincidence is characterized by the fact that two or more causal chains, which are completely independent of each other, coincide. Such a coincidental or accidental event therefore has more than one cause. Just before such an event occurs, when the independent causal chains coincide, "chance" turns into "necessity", using Jacques Monod's words. Essential coincidence is an event that actually occurred under given conditions, but, in comparison to other possible event chains, had an extremely low or even incalculable probability of occurring. It is a sudden change in the real world that can, in principle, be ascertained by observation or measurement, subsequent analysis of documents, traces, long-term consequences or the like. It can happen just like many other possible events, but is inherently unpredictable, was not predictable or, at least, not foreseen. Not least because the probability of such an accidental happening is close to zero. In retrospect, the causes of

---

[46] For comparison or for pleasure, you can look at various websites: https://plato.stanford.edu/entries/chance-randomness/ or https://link.springer.com/chapter/10.1007/978-3-319-26300-7_2.

[47] Cf. the following definition: "An event is referred to as happening by chance if it does not follow from a given set of conditions with inner necessity, if it could have happened this way, but also differently. This does not mean that a coincidental event is not causally determined. The universal validity of the causal principle also extends to accidental events." (translated by B. Wessling, originally in: Günter Kröber in Klaus, G., Buhr, M. (1969), Philosophical Dictionary. Bibliographisches Institut Leipzig. I would like to thank Dr. Rainer Feistel for this hint).

such unexpected or even unpredictable events can be understood and objectively analyzed,
even if sometimes not all causes are or can be recognized.

Many, if not most, processes in the world have several optional ways they can develop over time. We can often estimate the probability of different possibilities A, B, or C occurring; perhaps based on experience, or perhaps from theoretical considerations. But if a completely different possibility X occurs, which is extremely unlikely, and therefore no one will have considered it, we call that coincidence or accidental event, or: it happened by chance.

Let's take an everyday's example: Maybe we always go to the market on Fridays to do our shopping. With different probabilities, we will meet a neighbor A there (60% probability, we will meet him in 6 out of 10 cases), a sports buddy B (30%), and the mayor of our small town, whom we know (10%). Each meeting with these people at the market is a normal event, even the encounter with the mayor. It is not necessarily the case that we meet these people, but it happens often, which is quite normal, because they are also usually at the market at about the same time as we are on Fridays.

We would not expect to meet a former schoolmate there, someone we haven't seen in decades and have totally lost track of—but it happens one Friday anyway. Maybe because he was on a trip, his car broke down, he had the car quickly repaired at a nearby garage, and he took the opportunity for a walk into town. When he noticed the market, he wanted to stroll around there and maybe eat a bratwurst. He had no idea at all that we had ended up in this small town. It was purely by chance that we met again here, unpredictable, impossible to calculate a probability for this happening, and still, it happened: an essential coincidence. In retrospect, it can be understood, but not foreseen: We observe essential uncertainty.

Next, we look at the growth of a tree: It is 100% likely that the tree will form branches. But when exactly and where this happens, we can neither calculate nor predict. Branches can grow in completely different places, in completely different directions, the branching can begin at completely different times. These are simply essential coincidences. This results in impressive, unique structures of a tree crown, as we can in the photos below showing an Australian fig species (left) or this tree crown landscape in an Australian national park (right, photos by the author).

 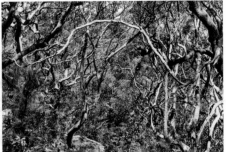

According to Jacques Monod, evolution is the prime example of essential chance: Normally, during fertilization and cell divisions, the DNA strands are read and copied correctly. A mistake happens only about once in a billion copying cycles.[48] In most cases, the error has no consequences, nothing noteworthy happens, or it is lethal or causes a disease that is disadvantageous to the individual. But in an extremely small number of cases—I don't know of any probability estimate—the mutation causes something positive. Even more rarely, after numerous other such accidental mutations, a new species emerges.

How rarely mutations lead to new viable species, we can already see from the reproduction rate of insects and common birds such as sparrows, which hatch in billions and billions, with mutations occurring correspondingly often ($1:10^9$). But how often does a new sparrow species arise? Or a new mayfly species? And when was the last new mouse species created? Mice have a very high reproduction rate; even during our own comparatively short lives there are hundreds of generations of mice in our immediate environment, dozens or hundreds of billions of mice worldwide. We cannot even give an approximate probability of when or how often a new member of this family of species could arise. It is extremely unlikely, but still possible —as we know from the study of fossils—even if we cannot elucidate the causes and circumstances that led to the emergence of a new species. And if it happens (it did during evolution, hundreds of millions of times) it is the result of a series of essential coincidences. The normal case, however, is that no new species arise from mutations.

It is important to note: **Coincidence, chance, accidental events do not mean that the corresponding events have no cause,** as, for example, the physicist, computer scientist, philosopher and non-fiction author Walter Hehl claims in his book *Chance in Physics, Computer Science and Philosophy* when he writes:[49] "Definition: Coincidence is a causal chain whose origin cannot be determined." First of all, coincidence is not a causal chain, but it *results* from a causal chain, more precisely: from several converging chains. Coincidence is an *event* that was caused by more than one causal chain. But look more closely at the quote from Hehl: "… whose origin *can not be* determined"—I cannot agree to that. It may be and is the case in many cases that we fail (or it is not worth the effort) to determine the origins of the causal chains. But that only such events, for

---

[48] http://mcb.berkeley.edu/courses/mcb110/Albertsch05.pdf, here reference [99].

[49] Walter Hehl, "Chance in Physics, Computer Science and Philosophy - Chance as the Foundation of the World", SpringerNature 2021, p. 23; original German edition "Der Zufall in Physik, Informatik und Philosophie—Zufall als Fundament der Welt", Springer Vieweg 2021, p. 23; on p. 38, in the subchapter 2.2.1 "Definition of Randomness", he writes, inter alia: "We define chance as a causal chain that starts without a past (as is the case with dice by definition) or as the meeting of two causal chains that, as far as they can be traced, have no connection." With the first half-sentence, I cannot agree, because, of course, every causal chain has a past, an origin. And dice do not cause essential coincidence/essential chance. With the second half-sentence (after "or") we meet on a mutual level of understanding.

which it is *principally not possible* to determine the origins, should count as coincidence is not in agreement with the property of essential coincidence. Because that means, in the definition of Walter Hehl: "… can not", that is: "It is impossible". For me, there is no question that the following statement is fundamentally wrong: "We speak of chance precisely because we have no causal explanation."[50] No, for most coincidental and accidental events, even we ordinary mortals can find a causal explanation, and for more complicated causal chains, researchers can decipher them. That may not always be easy, sometimes very difficult, sometimes it is not worth the effort. But there are always causal explanations.

Every event has a cause, usually several, and in essentially coincidental events, the event chains leading to the final causes originally have nothing to do with each other. According to my "description" of essential chance/coincidence above, there is at least one other much more probable event than the one we observe, at least one other possible event in the space of many possibilities in regard to how the causal chains could develop further, whether and when and how they intersect.[51] The French mathematician Poincaré formulated a definition of chance that I find a bit narrow, he writing "A very small cause, which escapes our notice, determines a considerable effect that we cannot fail to see, and then we say that the effect is due to chance."[52] These types of chance, which we will also look at in Chap. 6 under the term *Butterfly effect*, are the ones we want to look at in this book, even more so the ones that come about through the interaction of causes that are absolutely independent of each other, the essential coincidental events. Ultimately, the effect mentioned by Poincaré is an aspect of what we have described above: It is not a very small *cause,* which *causes* something big, but a mixture of unrelated causal chains that leads to an unstable situation, in which a tiny effect is then enough (which may escape our notice, as Poincaré wrote), to trigger a large event, an essential coincidental event. The drop that makes the barrel overflow. The examples I have described so far are all coincidental events of the essential kind. For each of the coincidentally occurring events, there was always at least one much more probable alternative. And extremely minor deviations in the course of the independent causal chains have been able to trigger, instead of the described event, nothing particularly exciting or a completely different accidental event. On that particular Wednesday, if I had missed the deadline for submitting my job ad, or if Mr. Petersen had been able to fall asleep easily, I would never have started working at Zipperling (and I would never have started this research which ultimately led to this book).

---

[50] Walter Hehl, ibid., p. 35. (in both editions, German and English, same page number)

[51] One can also say of such types of randomness that they are "contingent", cf. https://www.spektrum.de/lexikon/philosophie/kontingenz/1129.

[52] Quoted (in the original German edition) from Henning Genz, "Wie die Zeit in die Welt kam - Die Entstehung einer Illusion aus Ordnung und Chaos", Rowohlt paperback 2002, p. 297. english translation here cited after https://www.brainyquote.com/authors/henri-poincare-quotes

Alternative Views of Chance and Coincidence

A very comprehensive book has been published by Klaus Mainzer in German under the title *Der kreative Zufall. Wie das Neue in die Welt kommt* (C. H. Beck, 2007). The back cover states: "Chance proves to be the central theme of modern natural and social sciences". The author assumes chance as given and shows the role of chance in many different contexts: in early world views, in gambling, in probability and information theory, in connection with thermodynamics, in the quantum world, in life and in the brain or consciousness, in culture, economy and society. He also introduces the philosophy of chance. However, there is no discussion whatsoever about how the many chances arise, *why* there are chances at all. It just exists, we encounter it everywhere, we constantly see its effects. Klaus Mainzer is not alone in this, because almost nothing can be found about the causes of the emergence of chance (unless one is content with what is often asserted quickly, with one sentence, and without further justification: It is said to be the quanta, the elementary particles, that cause the chances everywhere; see the note in the next paragraph; we will think about this in more detail later).[53]

The founder of the Max Planck Institute for biophysical chemistry in Göttingen and Nobel laureate Manfred Eigen[54] and his colleague (later his second wife) Ruthild Winkler published a book in 1975 that was also internationally very successful, entitled *Das Spiel - Naturgesetze steuern den Zufall (English edition title: "Laws of the Game: How the Principles of Nature Govern Chance")*.[55] One might be tempted to believe that this book explains *how* the laws of nature generate chance and coincidence or accidental events. But this is not the case. As early as p. 35 in the introduction to Chap. 3 ("Microcosm—Macrocosm"), the authors address the elementary particles, the quanta, but also the atoms and molecules (citations from German text, translated by the author of this book): In this world, "all elementary processes are happening." They continue: "Chance has its origin in the indeterminacy of these elementary events. [...] Under special conditions, [...] there can be [...] a macroscopic representation of the indeterminacy of the microscopic 'dice game'." With the reference to the *dice game*, the authors point to the central theme of their book: They try to make the behaviour of, above all, evolutionary and other complex processes understandable and to *simulate* them using various parlour games and their rules, where, at the beginning of the game, it is not foreseeable who will be the winner and who the loser. There is nothing wrong with this, but a simulation is not an *explanation* of the actual mechanisms in nature, as it only describes the

---

[53] Klaus Mainzer, Der kreative Zufall. Wie das Neue in die Welt kommt; C. H. Beck 2007. (title translated: The creative chance. How the New comes into the world)

[54] https://www.mpinat.mpg.de/607453/Manfred_Eigen

[55] M. Eigen, R. Winkler, "Das Spiel - Naturgesetze steuern den Zufall"; Piper series 1975 (page references in the following from the German 1985 edition). English edition: "Laws of the Game: How the Principles of Nature Govern Chance", Princeton University Press 1993

outward appearance of the course of such processes more or less well. Dice games cannot answer the question of *how and why* chance and coincidence arise in nature on the basis of its laws. The indeterminacy of quantum processes (Heisenberg's "uncertainty principle") is certainly not the cause, as we will see in Chap. 6 of this book. And dice games are not based on *essential chances*. But in evolution (and not only there) we find countless such essential chances and coincidences, which are a characteristic feature of these natural processes: It is inherently uncertain for us what effect a certain mutation will have; when, where and if really a bolt of lightning will strike and which exact path it will take; what will be the stock value of a certain company or even what a national index will be in 12 months—examples of essential coincidental events. But the winner of a parlour game can be predicted with a certain and relatively high probability; a dice game has only six optional results, each parlour game has only a limited number of possible routes—nature has an unlimited number of options, whether we look at evolution, or at branching in trees, the paths of bolts of lightnings, the pattern of snow flakes, everywhere, we see the results of essential chance and coincidence.

David J. Hand published a book entitled *The Improbability Principle: Why Coincidences, Miracles, and Rare Events Happen Every Day*. Although the subtitle of the book promises to answer the WHY question (which demands an explanation of causes), this does exactly not happen. Even in the subchapter "Where does coincidence come from?" it is not explained where it comes from, but everything is described with probability considerations and mathematics—as it is throughout the book. In this respect, it is a very instructive book, while with mathematics, one can only *describe randomness*, but not essential coincidence. Moreover, one cannot understand *how* it comes about, and this has been happening for billions of years throughout the universe.[56]

Randomness is also discussed on Wikipedia.[57] On the equivalent German page, it allegedly represents events for which no causal explanation can be found. In the course of the text, various levels are mixed together. So, the question "Which processes are random, which causal?" causes more confusion, because, of course, unpredictably occur-

---

[56] David J. Hand, "The Improbability Principle: Why Coincidences, Miracles, and Rare Events Happen Every Day", Scientific American 2014, German edition "Die Macht des Unwahrscheinlichen", C. H. Beck 2015. Ian Stewart makes a similar argument, accompanied by a lot of statistics, in his book "Wetter, Viren und Wahrscheinlichkeit—Wie wir die Ungewissheiten des Lebens berechenbar machen", Rowohlt 2022. Original English edition "Do Dice Play God? The Mathematics of Uncertainty" Profile Books, London 2019. But the uncertainties of life, the coincidences and accidental events, are not calculable - as is alleged in the German subtitle: "How we make the uncertainties of life calculable" -, because the probability of their occurrence is close to zero, which can be seen with Ian Stewart's book as well: The original English edition was published a few months before the emergence of the Covid-19 pandemia, Stewart did not predict that it would happen, he had not pre-calculated this uncertainty of life; the German edition, published in 2022, contains a foreword about how pandemia are spreading ... .

[57] https://en.wikipedia.org/wiki/Randomness

ring events also have causes, and are therefore *causal,* only it is so that there are usually several other different and much more likely course options. And just as with Eigen/ Winkler, *randomness* in the microscopic and macroscopic world is practically thrown into one pot with the uncertainty in the world of quanta, elementary particles ("microscopic phenomena are objectively random", as the English wiki page tells us; however: elementary particles are not "microscopic", but far, far smaller if they have a size at all, and they do not behave randomly, but "uncertainly" or "indeterminably"). Monod also believes that the indeterminacy relation (uncertainty principle) of quantum mechanics is "on a microscopic level … [an] even more decisive cause of [macroscopic] indeterminacy"[58]—with all due respect, that is not the case. The different organizational levels of matter must not be mixed together. We will look at this in more detail later. In the world of elementary particles (on the "quantum level"), there are quite different laws than in the world of nano particles, on the next higher microscopic or, finally, the macroscopic systems. At the end of Chap. 6, we will discuss this under the keyword *decoherence.*

Two (German) authors, who have achieved great success as bestsellers of non-fiction books, also published popular science books about chance, coincidence, randomness. Both use statistics on the one hand and also the most basic level of matter, quantum mechanics, on the other hand, as explanations for the constant occurrence of chance and randomness. Randomness that can be investigated using statistics is not made up of essential chances, and quantum mechanics does not behave randomly, but *uncertainly* or *indeterminably.* The position and momentum of an elementary particle cannot be determined exactly at the same time, the measurement results are *fuzzy* (Heisenberg's uncertainty relation). Stefan Klein only briefly mentions in his book, "Alles Zufall—Die Kraft, die unser Leben bestimmt" (*Everything is Chance—The Power That Determines Our Lives*),[59] the reasons why, in his opinion, the events in the world are predominantly determined by chance or coincidence, and how chance came into the world. This is surprising because the title promises it and because a subchapter with the heading *How Chance Comes into the World* announces an explanation. However, it only describes which considerations lead to the conviction that the world is *deterministic,* that is, not ruled by chance at all. Essentially, the author describes a mathematical (statistical) interpretation of chance and, on this basis, discusses the effects and perception of chance in everyday life. It is also misleading to refer to chance as a "force", which continues to happen beyond the subtitle of the book.

Florian Aigner—like many physicists—makes the already described very fundamental error in his book *Der Zufall, das Universum und du—die Wissenschaft vom Glück*[60]: He throws chance, as we know it from our macroscopic world, together with the indeter-

---

[58] Jacques Monod, loc. cit., p. 108.

[59] Stefan Klein, "Alles Zufall - Die Kraft, die unser Leben bestimmt", Fischer Taschenbuch 2015.

[60] Florian Aigner, "Der Zufall, das Universum und du—die Wissenschaft vom Glück", Brandstätter 2016.

minacy of elementary particles into one pot. But these are two completely different and separate phenomena. The indeterminacy on the quantum level, which we cannot directly perceive, is not the cause of the existence of chance in the world that we do perceive (see also Chap. 6). Later in his book, finally, *Die Wissenschaft vom Glück* ("The Science of Happiness", the subtitle) is dealt with, mainly from the perspective of probability theory. But we also need to deal with the rather idealistic point of view that the author takes in his book. He writes: "Without chance there would be no us, and without us there would be no chance." By this, the author means, as he writes a little later, that there is only chance (or coincidence) because we can observe it, and thus name it. While I agree with the first half-sentence ("Without chance there would be no us"), I immediately detect a lack of logic in the second. Because, before there was us, there must have been at least one chance, at least one coincidence, more likely several more, from which we humans then emerged. But these chances could not have been observed by humans at the time. How then can coincidence only exist because there are people who observe it? Moreover, I consider his further statement, according to his own words based on the *anthropic principle*, to be rather far-fetched: "The world is built in such a way that it can bring us forth, and that is no chance." (or: this is not coincidence)[61] Five pages earlier, we read in his book "without chance there would be no us"—that does not fit at all! There, then speculations are following about universes that could be quite different than ours, and thus have not produced any humans. I consider such statements, at best, as not helpful, not conducive to knowledge, because they are neither verifiable nor falsifiable—thus, according to our understanding of science, they are unscientific.

But it goes on:[62] "We humans are not there because chance has brought us forth, but chance is there because we humans have brought it forth. One can only speak of a coincidental event if there is someone who experiences this event as coincidence." Again, this must be vehemently contradicted. Chance, essential coincidence, exists independently of whether *we* can *observe* it or not; our observation merely enables us to *recognize* it and to *designate* it as such. Coincidence, chance, accidental events have always existed, from the beginning of time, even when there were no humans yet. There were also coincidences that took place in the past when the humans had already appeared but had not yet developed philosophy, let alone natural science, and yet these humans were already capable of suspecting the effect or message of a supernatural power (thus nothing random) behind objectively accidental events (e.g., the impact of a lightning strike at an unpredictable location), which they then worshiped. And there were also all the chances and coincidences in the very distant prehistoric past that we are only now discovering through excavations, fossil finds or even—as recently—with the measurements of gravitational waves. The accidental event that two neutron stars, a neutron star and and a black hole, or two black holes collide has already taken place tens of thousands,

---

[61] F. Aigner loc. cit., p. 227.
[62] F. Aigner loc. cit., p. 228.

hundreds of thousands or millions of years ago, so it had already taken place long before we perceived and measured this event as coincidental.

The previously quoted physicist Walter Hehl has written and published a much more comprehensive book on "Zufall", on chance and coincidence.[63] As the title, *The Chance in Physics, Computer Science and Philosophy—Chance as the Foundation of the World*, promises, the author deals with physical, philosophical and, above all, information technology aspects of chance and coincidence. Nor are mathematical and biological considerations missing. Overall, the book provides a very good and far-reaching overview of the efforts to understand chance (or: coincidence)—including in terms of history of science and philosophy. In the course of our discussion, we will return again and again in the following chapters to individual passages from Hehl's explanations and critically assess them. We will not deal with the strong IT orientation of his chapters, in particular, not with the dubious argument that the brain is nothing other than a computer, or with the claim that evolution is to be understood as software technology.

According to the author, chance/coincidence occurs everywhere and has always been present in our world; it is—as mentioned in the subtitle—the foundation of our world. "Without chance (coincidence), the world would be unnatural, not nature. [...] Without chance, there would be no evolution."[64] Ultimately, he also believes that chances, coincidences and accidental events are caused by the properties of the quanta, of the elementary particles, something we will critically examine and discuss later in Chap. 6.

In his book, Hehl describes and discusses what he calls a "model coincidence", which exists when one jumps into a lake: "From the lake's point of view, it is a coincidence" (what happens to the water when the body dives in), "because there is no connection between the lake and my brain."[65] An analysis of the phenomenon of *coincidence* should not, in my opinion, be carried out "from the point of view" of an observer who has no means to "view" the event and to investigate the causes. This would mean that a subjective view of an objectively occurring event should be relevant for understanding the causes of chance. But that would be unscientific. The lake cannot have a *view*. If the lake were a conscious, living and analyzable being, it would discover that the human being has decided to jump. Maybe it would even notice that he does it from that spot more often whenever, for example, the weather is good and the water is warm and the human being has time. That the waves generated shortly after the jump have disappeared without a trace does not show that a "decay of chance" has taken place here, as Hehl writes: "Even the largest supercomputer is then no longer able to reconstruct a single sunken event like this jump. The causal chain, starting with the jump, has disappeared in the noise. Everything has been, in principle, computable and deterministic—but

---

[63] Walter Hehl, "The Chance in Physics, Computer Science and Philosophy—Chance as the Foundation of the World", Springer Vieweg 2021.

[64] W. Hehl, loc. cit., p. 128.

[65] Walter Hehl, loc. cit., S. 113 in the section "The Decay of Chance".

undetectable." I do not consider this to be an argument for a *decay of chance:* Most of all the chances, coincidental or accidental events that had ever taken place have probably led to consequences that can no longer be reconstructed after a certain time. Nobody (except the participants themselves) can reconstruct where, how and when two lovers *accidentally* met, unless it was meticulously written down or recorded on video. All the chances that have ever taken place on earth and will take place in the future will no longer be reconstructable when, at the latest, the sun will have swallowed the earth, and even less so when everything in our solar system will have disappeared into one or more black holes. That does not change the fact that they took place. These chances or coincidental events have not decayed, but only their traces. The information that was previously present in the traces has disappeared, and, accordingly, entropy has increased (more on that later).

In the course of this book, I try to convey a feeling for and first understanding of why essential coincidences can be *described* with probability theory, but can not be *explained* by either it nor by quantum theory. Just as unhelpful is the use of a term known as *absolute chance (coincidence)* for such phenomena, which are allegedly "fundamentally indeterminate and rationally inexplicable" ("indeterminism").[66] What are such phenomena or events? Who decides that something is or was *fundamentally indeterminate* and *rationally inexplicable*? Can it not be enough for a scientific and rational penetration of the world that something appears inexplicable to us at the moment, thus we do not claim to be able to explain something that we (at least currently) do not understand, instead of calling it *fundamentally indeterminate* (i. e., forever not explainable)?

Here,[67] and in other sources,[68] we once again find the assertion that the quantum world is both an example and a cause of *indeterminism*, an idea that, according to my understanding (see Chap. 6), is forbidden. It is very strange, in this context, to see *turbulence* designated as an example of indeterminism. [69] Hehl writes in a subchapter about turbulence: „[…] Indeterminism is the doctrine that an event came into being without a cause […] It would all be turbulence. […] If no flow thread can be traced from A to B, the flow […] is indeterministic." Of course, we are not able to follow every single water molecule in a turbulent flow; that would require an infinitely large effort (and influence the course of turbulence). But, nevertheless, turbulence is not an indeterministic phenomenon, because the behavior of the flow does not arise without cause! The famous physicist Richard Feynman called turbulence one of the last great unsolved problems of classical physics,[70] a statement from the 1960s that referred to the fact that there was still

---

[66] https://en.wikipedia.org/wiki/Indeterminism

[67] Walter Hehl, loc. cit., p. 34.

[68] https://www.spektrum.de/lexikon/philosophie/determinismus-ndeterminismus/438

[69] Walter Hehl, loc. cit., p. 104 f., in section 4.2.3 "Turbulence" and 4.2.4 "Between Chances and Order".

[70] Feynman R., Leighton R. B.& Sands M. 1964 „The Feynman lectures on physics" Boston, Addison-Wesley, cited after https://royalsocietypublishing.org/doi/10.1098/rsta.2010.0332, *siehe* https://physicstoday.scitation.org/doi/10.1063/1.3051743.

no parameter-free ("ab initio") theory of turbulence at that time. There were only mathematical generalizations of experimental results, which are approximations, *descriptions,* of the actual dynamics. However, in 2013, the theoretical physicist Helmut Baumert presented such a first parameter-free ("ab initio") theory,[71] in which no weaknesses have been found so far.

I consider metaphysics, which postulates supernatural forces for indeterminism, to be unscientific. I also assume that the question of the free will of humankind *cannot* be classified under "indeterminism", but rather belongs in my "coincidence" category, because, in principle, there is often one, if not more alternatives available to me for a decision I make: "Should I go for a walk or play the piano or take care of the more urgent bank matters now?" According to my understanding, all the thought processes that take place there are of a material nature. In addition, there are also indications that our supposedly free decisions are actually made by our brain beforehand, and only then become conscious to us[72]—that would be evidence for classifying willful decisions under *chance/ coincidence.* But it would be too much to also discuss the very complicated topic of *free will* here.

Let us therefore hold for the considerations that I present in this book: Chance and coincidence, as well as accidental events, are such events to which there are alternative event possibilities. And even with the best and most detailed knowledge of the conditions, we can predict neither what will happen and when, nor which of the alternative options, which we may not even know all or only in part, will be chosen by the relevant system. And it could even be the most unlikely or even the most unthinkable event option. Chance/coincidence is *essential* for such systems, as Monod called it.

Chance in Philosophy

I would like to deal with a special philosophical book. Its author is the Marxist philosopher and science theorist Herbert Hörz. In addition to numerous other publications, he has written an entire book on this topic entitled "Zufall—eine philosophische Untersuchung" (translated: *Chance—A Philosophical Investigation.*)[73] I chose this book

---

[71] H. Z. Baumert, „Universal equations and constants of turbulent motion", Physica Scripta Vol 2013, No. T155, paper 04100, vgl. https://iopscience.iop.org/article/10.1088/0031-8949/2013/T155/014001, also the somewhat more advanced presentation in the appendix, because this is a theory of the turbulence of Newtonian fluids; Helmut Baumert and the author of this book have jointly designed a first parameter-free theory also of the turbulence of non-Newtonian fluids.

[72] https://www.spektrum.de/lexikon/biologie/willensfreiheit/70771

[73] Herbert Hörz, "Zufall—eine philosophische Untersuchung", Akademie-Verlag Berlin 1980 (Schriften zur Philosophie und ihrer Geschichte, Vol. 24); digitized edition in the Max Stirner Archive accessible here (German): http://www.max-stirner-archiv-leipzig.de/dokumente/Hoerz_Zufall.pdf.

because the author explicitly refers to scientific findings. He formulates a criticism of Monod right from the start. It is about unrelated event series: This designation should not lead to a claim of their complete independence, because there is an objective connection between them due to the possibility of such accidental conditions in the way people live. Of course, I would like to answer, the accidental drop of the hammer on the doctor's head, or the equally coincidental discovery of my job application by Mr. Petersen in a newspaper section that he only looked at once in his life, was only *possible* because the lives of the doctor and the craftsman (or those of Mr. Petersen and myself) *allowed it*. The objective connection *arises* only in the moment when two previously completely independent event series meet by chance. Monod does not claim the existence of an "absolute chance" here, as Hörz writes, indeed, such a concept appears nowhere in Monod's book; instead he defines and explains the concept of *essential chance (or: coincidence)*.

In the foreword to the digitized edition (2013), which can be understood as an update of the older book, the author summarizes the conclusion of his book in five points. One aspect is the dialectic of "chance and law" (or "necessity"), that is, an idea practically corresponding to the title of Monod's book. This formulation is very convincing: "We recognize the […] coincidental realization of a possibility from the possibility field contained in a law." This describes very well that, with chances, also with the *essential* ones, which is what we are examining in this book, it always concerns processes, which are also *possible* according to the natural laws.[74]

Another aspect, in his opinion, is the role of probability theory: "Chances are events that can be […] realized with a certain probability. […] Probability theory recognizes laws […] and statistics provides material for this. Chances are recognizable in this sense. […] The control of chance includes the aspect of knowledge, planning and activity. Risks are to be recognized. Risk prevention is to be taken." On p. 17, Hörz underlines his above-mentioned criticism of Monod with the fact that statistics would make statements about how often such accidents (of course, not exactly the one described by Monod) would occur, and that insurers would take the results of these calculations into account in their contracts.

This corresponds to widely held and largely accepted views among natural scientists and philosophers, but I do not agree. Because, on the one hand, the really decisive chances and coincidences are unique. I have already given enough examples illustrating that essential coincidental events occur with an infinitesimally low probability and are not predictable. Fukushima has already shown where the limits of probability theory (and the insurance industry) lie in risk prevention. On the other hand, probability the-

---

[74] His definition of chance is as follows (ibid., p. 71): "Chance is an objective relationship between the inexhaustible properties of an object, a process or a person (group) and between the inexhaustible relationships of different events, which does not result from the essential internal conditions of these related components."

ory can, at best, *describe that,* but it can *not explain to us why* there are so often such decisive, essential chances in our world, or why this or that particular chance occurred. Statistics can also not give us an answer to the question of how often such a case as Petersen's first and only reading of the academic job advertisements in the FAZ science section, in which my advertisement was to be found, occurs, because the probability is infinitesimally close to zero. Hörz also ultimately sees the cause of all chances in the properties of elementary particles.

The term *absolute coincidence,* as well as *true* or the already mentioned *objective* coincidence, is repeatedly thrown into the debate. What could this be in contrast to *essential chance (coincidence),* as Monod defined it and as I described it in more detail above? In contrast to an *absolute,* is there also a *relative* chance, in comparison to the *objective* also the *subjective,* and, if so, what should each be? These terms suggest to me unclearness and confusion rather than fundamental understanding.

According to Charles Sanders Pierce, "the absolute chance plays an active role in the universe"; the doctrine that he designed is called "Tychism".[75] Within the philosophical category *indeterminism*, the term "objective chance" denominates such accidental events, which (in contrast to the subjective chance) are not reducible, that is, not dependent on (hidden) causes, but fundamentally indeterminate and rationally inexplicable. This type of chance is again said to be caused by quantum processes.[76] The reference to the elementary particles with their indeterminacy, which is supposed to cause macroscopic coincidence, does not help us further, as we will see in Chap. 6.

Coincidences are Real and Everywhere and Anything but Rare

Coincidences are an objective reality regardless of whether we observe or recognize them and perceive them as such. There are (from our point of view) beautiful and ugly and even outright bad coincidental events. And very often, it goes on like for the farmer Sai Weng: You don't know what will develop from it. It is purely by chance that the earth is in a habitable zone, and even this chance would not have been enough to make life relatively stable, as we know it here. It happened by chance that the moon—probably after a gigantic cosmic collision—was formed, and that the direction of the earth's axis was stabilized and seasons resulted. If that had not been the case, climate zones would not have been able to form, a development that—despite all the climate changes over time—allowed for a slow evolution. Instead, the earth's axis would have undergone chaotic fluctuations.[77] We will not deal with this any further here, but we do want to deal

---

[75] Walter Hehl, loc. cit., p. 249.

[76] https://en.wikipedia.org/wiki/Indeterminism, see there footnotes [1] and [2].

[77] J. Laskar, F. Joutel & P. Robutel, Stabilization of the Earth's obliquity by the Moon, Nature Vol. 361, pages 615–617 (1993) https://www.nature.com/articles/361615a0

with events that occured long after the accidental formation of the moon, namely, when life, and evolution with it, had already begun. And after life had begun, it would not have lasted for three or four billion years if the earth, by chance, had not had a number of mechanisms with which the climate could be kept relatively stable over these long periods of time. Simulations show that planets without such *thermostat functions* practically have no chance of developing higher forms of life, because they would either freeze or boil over.[78] For this alone, we should bear in mind that the sun is currently radiating thirty percent more strongly than at the beginning of life, and thus the temperature regulation of our planet does indeed work excellently.

The Beginning of Life and Evolution

Too little is known so far about how life itself began. There are a lot of serious hypotheses about how it may have finally come about that there were single cells whose properties were genetically passed on by doubling of the DNA, and that this also happened at some point by means of sexual reproduction. It could be that, at the beginning, there were only simple catalytic reactions that led to short protein chains, perhaps in lipid-coated droplets, or amino acids could have stabilized the first cell precursors.[79] Or was it only RNA that started self-replication, followed by mixtures of DNA and RNA?[80] Nobel laureate Manfred Eigen designed a *chemical evolution* in *Hypercycles* as a model for evolution in a preliminary stage.[81] But we can be quite sure that chance, coincidence and accidental events played a key role in the emergence of life.[82]

When the first single cells formed and then sexual reproduction developed, evolution had created a tool with which life could become more and more diverse over millions and billions of years: mutation. Species arose that could not only tolerate changing

---

[78] https://www.sciencealert.com/are-we-just-lucky-that-earth-has-stayed-habitable-for-billions-of-years *or* https://www.inverse.com/science/earth-has-stayed-habitable-for-billions-of-years-exactly-how-lucky-did-we-get.

[79] https://www.sciencedaily.com/releases/2018/08/180830180101.htm

[80] https://www.sciencedaily.com/releases/2020/12/201228095428.htm

[81] https://www.spektrum.de/lexikon/biologie-kompakt/hyperzyklus/5806, (German) see also M. Eigen, P. Schuster, "The Hypercycle—A Principle of Natural Self-Organization", Springer 1979; see his condensed article version M. Eigen, "Molecular self-organization and the early stages of evolution", published online by Cambridge University Press, 2009, https://www.cambridge.org/core/journals/quarterly-reviews-of-biophysics/article/abs/molecular-selforganization-and-the-early-stages-of-evolution1/3E87152BDEB279EE51A00CAB69748C0A

[82] As early as the 1980s, Robert Shapiro pointed to the possible and decisive role of RNA and, above all, to the role of chance in the origin of life, see Robert Shapiro, "Schöpfung und Zufall" C. Bertelsmann 1987 (German). Original English edition: "Origins: A Skeptic's Guide to the Creation of Life on Earth", Summit Books/Simon & Schuster 1986

climatic conditions, but actually needed them to reproduce massively, and that entered into relationships with each other, and thus formed the most diverse ecosystems.

Let's take a look at the mutations that made this possible. When DNA is doubled, there are reading errors; sometimes, part of the triplet code (a "letter"), which stands for a certain amino acid, is missing. Sometimes, the *right* complementary base is not incorporated, but a *wrong* one, a different one. This then leads to the fact that, during enzyme synthesis controlled by this gene, a different amino acid is read out or even a completely different sequence is created. Because, if a letter is missing, the triplets are suddenly read differently: for example, from AAA-BBB-CCC-DDD- etc. to AAA-BBC-CCD-DDE- etc.. This will probably lead to nonsense, that is, to a protein that does not work as an enzyme. And, as has been found, most mutations that are not lethal are meaningless. But some are not.

The translation of DNA into RNA into enzyme synthesis works according to the same rules in all organisms. So, the genetic code and its implementation in enzymes have not changed in billions of years.[83] We can assume that there were alternatives to this in the beginning, but, probably for reasons of efficiency and error tolerance, the system that we know today then prevailed, and it is the only program used with which heredity, growth, metabolism, etc., are controlled. Although there are deviations in the code meanings in some species, and this probably mainly in the *cellular power plants,* the so-called mitochondria, where the energy supplied from outside is converted into a form that is usable in the cell. But this only shows us that there were different ways to the standard program, in addition to which it shows that the standard program itself is subject to evolution. Anyway, the result is that there are enough errors by chance to fuel evolution, but at such a low level that it has an incredibly high error tolerance. Yes, even a repair mechanism is implemented!

The result is a system that only allows very few accidents, and these fall under our definition of *essential coincidence.* Not because they deeply affect life—of course they do!—but because they are essential to the essence of heredity, to how heredity is operating: DNA replication cannot and must not be error-free, but it is extremely error-tolerant (one exchange of a base for a *wrong* one per one billion copying cycles).[84]

In terms of the question of this book ("How do chance and coincidence come into our world?"), we should now remember that not only is the mutation itself, i.e., the change of information on the nucleotide chain, an accidental event, but, of course, also the resulting structure and even more so the changed evolution-relevant function. Those who hold that indeterminacy on the level of elementary particles, of quantum particles, are responsible for the causes of chance in our macroscopic world as well, often and gladly cite that the behavior of elementary particles is also responsible for the mutations. I am not aware of any evidence for this. Due to the phenomenon of *decoherence* (see Chap. 6), this seems impossible to me.

---

[83] https://www.scientificamerican.com/article/evolution-encoded/

[84] http://mcb.berkeley.edu/courses/mcb110/Albertsch05.pdf

Even more crucially, it is important to realize that these three coincidences can not be reduced to each other at all: the effect of coincidental event 2 (changed DNA information) cannot be derived from accidental event 1 (reading error = mutation); most especially the effect of coincidental event 3 (new function of an enzyme?) on the fitness of the corresponding species cannot be derived from the accidental error 1. This also makes it clear that *the information content* of the three-dimensional structure of the enzyme (and of the entire organism in which the changed enzyme is acting) is therefore *much higher* than the content of the corresponding sequence in the DNA! With this conclusion or insight, we are already anticipating a later chapter in which some important arguments are laid out about the relationship between different levels of aggregation of the—living as well as non-living—matter and their respective laws of nature, how they arise and how they interact with each other and whether they depend on each other or not. It is also a reference to later explanations when I add: The higher information content on the next higher organizational level above the DNA (reading of the DNA → construction of an enzyme → folding into a three-dimensional structure → function of the enzyme in the organism) is equivalent to a *lower* entropy content. Entropy has therefore been exported from the cell. We will talk about entropy in more detail in Chap. 4.

A new species does not arise—at least, in higher organisms—after a single mutation, but only over the course of many generations with many equally accidental mutations, which lead to further coincidentally changed or new proteins with other functions. And even if we knew the (positive) function of the first, and then also the (helpful) functions of the further changed or new proteins, we would still not know for a long time how this new organism (the new species) could prove itself in its environment. This environment is also constantly changing due to coincidental events of the weather, the climate, the changes in the river courses, the plant community, to name just a few factors.

The *neutral theory of molecular evolution*[85] has become a central part of the modern understanding of evolution. According to this, coincidental events such as *genetic drift*[86] play an important role in evolution, not just selection.

This makes it clear to us that the emergence of even one new species depends upon a sequence of several, if not many, completely independent accidental events. Chance/coincidence is therefore clearly an essential characteristic of evolution, already at the starting point, a mutation in DNA. Independent of this is the development of the function of a new protein, as is the development of the effect of this new protein in an already existing complex organism. ("Independent of" meaning: you can not predict from the mutation chemistry what the function of an eventually resulting new protein or new enzyme will be.) And then, chance intervenes again in the development of the properties of a possibly emerging new species, which must also assert itself in the again

---

[85] https://www.cambridge.org/core/books/neutral-theory-of-molecular-evolution/0FF60E9F47915B17FFA2620C49400632

[86] https://en.wikipedia.org/wiki/Genetic_drift

coincidentally changing ecosystems. Even the complex changes in ecosystem properties are essentially based on the action of accidental events. Here, I would like to call out to Florian Aigner[87] once again: All these chances and coincidences happened on earth many millions and even a few billion years before *homo sapiens* began to perceive and name them. There are simply and purely coincidental events independent of man.

Curiosities Cabinet of Evolutionary Coincidences

Evolution has constantly more surprises in store for us, surprises that are the result of an extremely long chain of mutations over the course of millions of years. Nobody can come up with such a coincidental result of evolution as this: There are fungi that take root in the bodies of insects and manipulate them so that the fungus can spread; the fungus *Orphiocordyceps unilateralis* does this with the carpenter ant; after infection, the ant climbs onto a plant, which it would never do otherwise; at an appropriate time, the fungus forces the ant to cling to the plant with a bite. Now, the mycelium of the fungus grows out of the ant's feet and attaches to the plant. The fungus digests the body of the ant, the stem sprouts out of the ant's head, and the spores spread down from above onto the ants crawling around there; the spores that miss their targets form sticky secondary spores that are practically traps on the ground, in which more ants get caught and then infected ...[88] And this is just one example of the results of countless strange coincidences of evolution, because neither did it develop deliberately like this, nor can it be planned, nor could anyone have foreseen it. It surprised even the most experienced experts when this was discovered by chance; no one would deliberately search for something like this, because it lies completely beyond our imagination, even the imagination of evolutionary biologists.

Many other examples of very strange products of evolution are to be understood in the same way. Take, for example, the virus that inflates its victims—an archaic single-celled organism that lives in volcanic springs—to eight thousand times their original volume, and thus creates a huge virus factory.[89] But the single-celled organism can defend itself by giving its offspring antiviral genetic material with the help of the CRISPR/Cas gene scissors, so that they are protected from infection.

---

[87] Florian Aigner, "Der Zufall, das Universum und du—die Wissenschaft vom Glück", Brandstätter 2016.

[88] Summarized and shortened from Merlin Sheldrake, "Entangled Life", Random House 2020 (the cited information was taken from the German edition: "Verwobenes Leben", Ullstein 2020, p. 148).

[89] https://www.pnas.org/doi/full/10.1073/pnas.2022578118

A white fly, the tobacco whitefly, a major plant pest, has become resistant to plant toxins by adopting genes from plants—probably via viruses.[90] This is only the second time that the incorporation of plant genes into an insect genome has been discovered.

Nature and Evolution is Aimless, *homo sapiens* is not the Summit of Evolution

Of course, views are always expressed according to which such developments in evolution are, as it were, purposeful, in the direction of optimal characteristics. Then, anthropocentric thinking culminates in saying that *homo sapiens is the crown of creation*, and thus the goal of evolution, its summit. But many evolutionary biologists and many observers of the findings of paleobiology agree: If the huge plant-eating dinosaurs and the predatory dinosaurs had not become extinct for any reason, the mammals, and thus the primates and then humanity would not have developed in this way. Nobody knows where evolution would have gone then.

What is more: There would be no humans in the form that we know them or think we know them if it were not for a mutation that took place about ten million years ago: As a result of this, a certain enzyme, alcohol dehydrogenase, became forty times more efficient in breaking down alcohol. Alcohol was poisonous already in small quantities before this. The mutation occurred in one of the last common ancestors of the primates.[91] In the end, this—long before humans evolved—provided the basis for humans to discover yeast, with which alcoholic beverages can be made from fruits or barley. This contributed to sedentary living and promoted spiritualism: shamans, medicine men, divining and prophetic people often stimulated themselves with alcohol and other drugs. However, yeast not only allowed for the production of alcohol, but also the production of bread, an important food in connection with the development of agriculture. So, the change from hunter-gatherer societies to farming may have been decisively supported by a mutation ten million years earlier. One could say that, without the mutation of alcohol dehydrogenase, there would have been no human culture, no formation of civilizations. Too many indications suggest that the production of beer and wine was already common in the oldest civilizations, not to mention that sedentary living in the form of village communities and growing populations caused a problem in regard to the hygiene of drinking water—but here, too, alcohol helped with its disinfectant effect. Without an efficient alcohol-degrading enzyme, this would not have been possible.

But this mutation alone would not have produced humans; it would have remained with primates who enjoy snacking on fermented fruits from time to time. The evolution

---

[90] https://www.sciencedirect.com/science/article/abs/pii/S0092867421001641

[91] Carigan et al., Hominids adapted to metabolize ethanol long before human-directed fermentation https://www.pnas.org/content/112/2/458.

of a primate line into humans required two more coincidental mutations, namely, in genes that arose in the human lineage after the branch that led to chimpanzees and bonobos split off.[92]

First, there is the duplication of a part of a gene responsible for neuronal growth in the brain. Such complete gene copies can occur from time to time, but do not necessarily cause anything noteworthy. In this case, however, it led to increased growth, the brain became larger! Second: at one point in this gene, a "letter" of the genetic code was misread. A "C" (cytosine) was replaced by a "G" (guanine), a tiny mistake. In this case, the *typo* caused something decisive: the characteristic furrow of the cerebrum emerged.

At this point, one might want to remark that all this is nice and good and right, but whether evolution leads to intelligent organisms like us humans or to other intelligent beings, the conclusion would be the same: In any case, evolution leads *purposefully* to *intelligence*. Really? I know that the certainly not unlikely assumption that there is also life on other planets in other solar systems of the Milky Way and even in other galaxies has many supporters. And I know that it is just as often speculated that there would also have been enough time there to develop intelligent life. Extraterrestrial intelligences could be technologically much further along than we are. These are all beautiful daydreams and mind games; I, too, enjoy reading about such things very much. Nevertheless, it will be very difficult for us to prove or even refute this. I am therefore asking the question differently: How do we come to the opinion that evolution—on the one hand on Earth, on the other hand on distant habitable planets—necessarily leads to intelligent life forms?

As an attentive observer and scientist, I cannot refrain from making a side note on the subject of intelligence before we get to the answer to this question. I do not want to join the choir of those who deny intelligence to humanity. But, at the very least, we have not yet proven to be intelligent enough to develop a sustainable (lasting at least some ten or some hundred thousand years) civilized and highly technological way of life, looking at the civilizations that we have built up and regularly destroyed again in the course of the last few thousand years. It currently looks more like we are not. Especially when you look at our really very efficient way of reducing biodiversity, one ecosystem after the other, and thus destroying our livelihood. And when we look at how little human life and cultural diversity count in military conflicts. We currently see this again in a particularly drastic way in Russia's attack on Ukraine.

---

[92] W. Huttner et al., cf various articles provided here: https://link.springer.com/article/10.1007/s12268-020-1411-5

Is Evolution Repeatable?

How can we then turn to the question of whether an evolution, be it on Earth or on other planets in other solar systems, necessarily leads to higher intelligence? Convincing answers can be provided by research when it investigates whether the chances of evolution are basically repeatable. There is the phenomenon of *convergent evolution* (e.g., the development of fins in fish and whales, that is, in the case of body parts that have the same function but have evolved from very different body parts in terms of evolution).[93] This could lead us to believe that certain conditions of the habitats necessarily lead the evolution in certain directions, that is, also in the direction of intelligence. Most animal species, including birds, are already quite intelligent, but we humans have a special kind of intelligence.

Fortunately, there is some systematic and usable research on this type of question. Richard Lenski started a long-term experiment to test it: "Would the biological evolution repeat itself in the same way again?" This task is, expressed in other words, identical to the question of whether, under corresponding ecological conditions, certain properties of organisms necessarily develop in the same way as, for example, fins; this is practically identical to the question of whether *intelligence* must *necessarily* evolve over the course of evolution. The answer by Zachary Blount, who works on Lenski's experiment, is: The evolution would not repeat itself in the same way again, "because coincidental events in the past influence further development".[94] We should keep in mind that renowned paleontologists and evolutionary biologists are of the opinion: If the evolution were to start all over again, the same biological solutions would come about again. The renowned evolutionary biologist Stephen J. Gould, on the other hand, holds the opinion[95] that, due to small deviations in the course of evolution, a completely different living world would eventually come about. I interpret this to mean that the development of intelligent beings like us humans is by no means necessary. It rather seems to me that we are a whim of nature, similar to the dinosaurs, our fascination with which is not without reason. Species and species families other than we humans have a much longer and much more stable evolutionary history behind them.

How did Lenski proceed in his research? He began his long-term experiment in 1988. Escherichia coli bacteria are kept on nutrient solutions under optimal conditions, and within 24 hours, about six new generations develop, totaling about 500 new generations every 75 days. The only thing that can happen here are mutations, while the

---

[93] https://en.wikipedia.org/wiki/Convergent_evolution.

[94] Zachary Blount, Richard Lenski, Jonathan Losos "Contingency and determinism in evolution: Replaying life's tape", Science. 2018 Nov 9;362(6415), https://pubmed.ncbi.nlm.nih.gov/30409860/, citations in the book text taken Blount's article in https://www.spektrum.de/magazin/der-zufall-in-der-evolution/1535073 and translated.

[95] Stephen J. Gould, Wonderful Life. The Burgess Shale and the Nature of History. Norton, New York NY u. a. 1989, German edition "Zufall Mensch", Carl Hanser 1991.

environmental conditions do not change. In this experiment, twelve (originally identical) bacterial lines are kept in parallel. This means that, if evolution under the same conditions would necessarily proceed in the same direction and lead to the same results, the twelve originally identical populations would show identical evolutionary trajectories after now more than seventy thousand generations under identical conditions. This number of generations would correspond to 1.75 million years in human development, at the time of the "Homo ergaster" in the development of Homo sapiens. Until Homo sapiens, who only appeared about 300,000 years ago, there were a lot of mutations in the course of another 1.45 million years, under drastically changing environmental conditions. But Lenski's experiment deliberately keeps the environmental conditions constant, so that they can not exert uncontrolled influences on the probability and effect of possible mutations.

But the bacterial populations are not identical at all after the aforementioned more than seventy thousand generations under *identical* environmental conditions. Without going into the many different interesting details: In some of the vessels, two or even more separate bacterial lines had developed, although they still looked quite similar on the outside (but the DNA sequence analysis showed the fanning out). However, only one *ecosystem* displayed a visibly different appearance: It was milky, not clear like the others, so it contained many more bacterial cells. This population had started to feed on citrate (which is present in the vessels) after 33,000 generations. Normally, bacteria can only use this in the absence of oxygen; here, this mutation occurred and took over the regiment in the presence of oxygen, which gave the mutated bacteria a selective advantage. Because they now had more food available compared to the bacteria that could only digest glucose.

After careful examination, it turned out that it was not one, but several mutations that led to this new ability. In other breeding lines, part of the previous mutations had also occurred, but the third decisive mutation had only occurred in one of the 12 identical ecosystems.

The conclusion from the previous tens of thousands of generations of the various evolutionary processes is: In general, the evolutionary directions can be similar, but the details differ because the respective different past always plays a role. And sometimes, something completely new, possibly unexpected, arises as a result of this past or "out of nowhere". It is very important to note that evolution does not look *forward at all,* that is, it is not goal-oriented. That is why we can also conclude: It is by no means inevitable that evolution leads to highly intelligent beings with the ability to abstract, do research and philosophise. It depends on chance, more precisely: on an inherently, essentially unpredictable chain of coincidences.[96]

---

[96]This is essentially also the content or the consequence of the so-called contingency theory, cf. https://en.wikipedia.org/wiki/Wonderful_life_theory. The opposite theory to this, which is apparently supported by a majority of biologists, is the "convergence theory", according to which certain properties of organisms (such as wings, fins, eyes) develop necessarily. By "contingent", one understands "something that is possible", cf. footnote 51.

Chance, Coincidence and Accidental events Are Ruling the World, in the Small as in the Large

We have now encountered enough examples of genuine *essential coincidences,* fundamentally different from randomness in roulette or dice games. Perhaps you now join my opinion: Chance, coincidence and accidental events rule the world. All of us, as well as the entire world we live in, are playthings of constantly occurring coincidences. We just have to determine, amidst the turbulence of unpredictable events, how we can best sail our lives, our professiosn, our companies, our scientific research on this incalculable stormy sea. Let me return to the very instructive book by Daniel Kahneman: "We generally overestimate our knowledge of the world and underestimate the role that chance plays in events. Excessive confidence in the predictability of the world is supported by the illusory certainty of retrospective insights."[97]

If chance and coincidence are the norm on Earth, in the solar system, in the universe, and not predictability and calculability, if predictability and calculability are rather exceptions: Why is that so? What are the causes? In my research, without planning it, I approached the answer to this question and would now like to invite you to follow me step by step on the way to understanding the complex causes of chance, coincidence and accidental events. In doing so, we will also take up the findings of many other researchers. These are findings that have gradually opened up to me over forty years, but on very convoluted, anything but straight paths. With my book, I try to shorten the way for my readers considerably.

---

[97] D. Kahneman, loc. cit, p. 26 (German edition, translated by B. Wessling), on p. 270, one finds another conclusion: "The thesis that significant historical events are due purely to chance/coincidence is deeply disturbing, but verifiably true." As a small example, he cites the fertilization of three female eggs by male sperm, which led to the men Hitler, Stalin and Mao TseTung. Each of these eggs could have led to a female child, and history would have been different. That all three were born as men had only a probability of 1:8. This leads to "the assumption that long-term developments are predictable, ad absurdum".

A small, but beautiful and unforgettable meeting by chance: I was walking along a narrow, densely overgrown path from which I had never been able to observe cranes (I am a crane researcher in my spare time). On this day, I happened to have my large camera, which I rarely take with me, and encountered this crane only twelve meters away from me. It was just three months old. I had never been so close to a wild young crane before. Cranes usually have a flight distance of at least 200 meters from humans. We looked at each other quietly for almost two minutes, me with the camera at my eye. (Photo by the author, July 2021).

Creativity is Coincidence in the Brain

3

Abstract

Coincidental events are constantly taking place in our brains as well. There are serious scientific findings that suggest "chance generators in the brain". Many people, some of them famous, achieved basic findings, discoveries and inventions as a result of a dream or situations in which they did not think consciously about the problem to be solved. The author was also able to solve the mystery of a previously unknown phenomenon in a situation in which systematic logical thinking was not possible. The phenomenon of improvisation, known especially from jazz music, is verifiably possible only if and when brain control mechanisms are turned off like while dreaming, so that chance has free rein.

Actually, I wanted to become a biochemist, but instead, I went to a small engineering company after university, and then I ended up working for an almost insolvent company that mixed plastics with other polymers and with additives. But why didn't I go to the USA or to Great Britain after my Ph.D., countries where biotechnology was slowly developing, where startups were setting out to revolutionize the pharmaceutical market scientifically, technologically and from a business perspective? Well, quite simply: I knew nothing about it. I was very provincial, my horizon didn't extend further than the borders of the Ruhr region, maybe it reached as far as the Münsterland in the north and the Sauerland in the south. About twenty years later, I briefly regretted not having changed to another city during my studies, let alone going abroad. Of course, I am responsible for that myself, but neither my parents nor my professors nor my doctoral supervisor asked me even once: "Why don't you go overseas for a while?" Because that's where the center of biochemistry and biotechnology was.

But over the years, I gradually learned that I had not missed anything. A series of coincidental events and my decisions at the forks in the road have ensured that I have not ended up in a cul-de-sac. If I had gone to the USA, I would not have invented the organic metal.[1] I would not have done crane research in my spare time,[2] an endeavor that has enabled me to gain so many additional insights into the world and life, not to mention stress relief, distraction, and the beneficial effects of frequent stays in nature. I would also not have spent thirteen years in China, where I definitely did not want to go, but which made for an unforgettable interesting, exciting and instructive phase in my life.[3] When the business coincidences overwhelmed me (customers from Germany and the USA shifted their production to China, and there were constantly huge technical problems and complaints there), I changed my *life plan,* which I had not previously had, overcame my fears and went to China to tackle the problems myself. I very quickly found a very good Chinese employee just by chance, and then began to fight my way through the quite different problems there. I started to learn and understand Chinese. I looked for a soccer team with whom I could play soccer on weekends, letting off steam and getting to know Chinese people outside of business life, and was soon called up to a second and a third team. I thought I would come home again after maybe two years, then it was five, and when finally the business had gotten better and better, I sold my chemical company, founded a small one-man technology consulting firm in ShenZhen and continued the business as a consultant. So, it became eight, and then thirteen years. By the last years, I was able to speak Chinese fluently (albeit by no means perfectly), and also to discuss and even to give lectures freely in Chinese on technical, scientific and business topics. I have been able to get to know the country a little, because I spent a lot of time in many towns in China, and also (but only in later years) in remote countryside areas in the North and the West of China, when I undertook several longer touristic trips. My horizon has expanded enormously. I have learned an incredible amount about this foreign country and its culture, and achieved unexpectedly great success in the market.

If none of that had happened, if I had gone to the USA as an advanced student or postdoc, I might have become the marketing manager of a biotechnology company in California, would have long since stopped doing research, might have had a heart attack at the age of fifty, and would have become an embittered green-card holder in the grey Trumpistan, another casualty, robbed of my ability to trust of the Trump and the post-Trump era.

But back then, before all these coincidences and accidental events, I was like the frog who *sits in the well and looks at the sky*, 坐井观天, (zuò jǐng guān tiān), as this Cheng Yu, on whom this story is based, put it:[4]

---

[1] It is important that I tell you about this novel material and the coincidences that led me to its discovery, because my scientific research that underlies this substance ultimately also led me to the understanding of chance and coincidence, which I would like to bring you closer to in this book.

[2] B. Wessling, "Der Ruf der Kraniche" (Goldmann 2020), English edition "The Call of the Cranes", (Springer Nature 2022).

[3] B. Wessling, "Der Sprung ins kalte Wasser", Verlagsgruppe Eulenspiegel, planned summer 2023.

[4] Text from the collection of proverbs of the East Asia Institute https://bit.ly/3HQwhTj, reprinted with kind permission of the institute.

In a dilapidated and derelict well lived a happy frog. He wallowed in the mud and slept in the holes in the well wall. One day, when a turtle came from the sea and asked from above how it was down there, he raved about his well: "It's wonderful here! Come and be a guest in my realm, where I am king."

But when the turtle wanted to put her right foot in the well, her left foot got stuck on a wall ledge and she hesitated. Then, she told the frog about the sea: "Do you know the sea? You have no idea how gigantic it is. Even if you imagine a distance of a thousand miles, you have no idea of the vastness of the sea—and even if you imagine a thousand meters, you have no idea of its depth. A long time ago, there was a nine-year flood—but the sea level did not rise. Later, there was a seven-year drought—but the sea level did not fall."

The well frog was quiet, didn't understand, didn't believe it, and the turtle went on her way.

This proverb is used to describe people with limited horizons who are not aware of it, and while it may be thousands of years old, it describes me well enough in my time at university to have been written specifically for me. The story exists in several variants, including one with a bird that tells the frog something about the world outside, which the frog does not believe, and also in this beautiful variant, which fits me even better in my younger years:

In a deep well lived a colony of frogs. They were protected from the outside world by their isolation. The only danger they were occasionally exposed to was the water bucket. Now, one day, a young frog named Froggy (蛙蛙, Wāwā) came up with the idea of using this bucket for a trip up. He spoke about it and received harsh criticism: "The sky will punish us; the world outside the well is terrible!" But our curious Froggy was not discouraged: With the next bucket, he traveled into the unknown. As a result, the Frog King (蛙王, Wāwáng) banned any mention of this incident.

Months passed, and Froggy continued to be on everyone's minds. Then, one day, a familiar croaking was heard from above. Froggy sat on the edge of the well, his wife and seven small frogs next to him. "There is a wonderful world up here," he croaked into the well. The Frog King threatened: "Woe to you, you traitor! You will be put to death if you dare to come home." But Froggy only replied disdainfully: "I will never come back to your dark hole. Have a nice day."

Now, there was an uproar in the well, which the Frog King ended with violence. But the next day, when the bucket was pulled up again, it was full of frogs.

It was not until my second professional employment that my view changed, my horizon expanded, and I wanted to get out of the well. So, I became the little "WaWa" who used the bucket for a trip up and stayed there in the unknown, but wonderful world. I learned that, for a certain period of time, Zipperling had had a cooperation agreement with an American company, when the former had begun to develop color concentrates for plastics. This cooperation had eventually been terminated, as Zipperling then felt competent enough, and the exchange with the Americans was considered financially no longer sustainable. But now, I suggested starting a different kind of cooperation with this company from Ohio. I flew to the USA to meet the local experts and made a proposal: "Let's cooperate strategically, stimulate each other, help each other with solutions for the development of products, without billing each other. We are not going to the USA to look for customers there, you are not coming to Europe. We can talk about how to cooperate in Asia later."

The shareholders of the local company and their various department directors agreed with my proposal, and I found a friend in their development manager. We met three to four times a year, either in the USA or in Germany. This was the beginning of my international business life, which was later enriched by many international research contacts.

Cooking is Good Craftsmanship, Science Must Go into Depth

But I was increasingly bothered by the fact that we developed everything purely empirically. Basically, we worked like cooks; okay, like ever better cooks, but in essence no different. Maybe I was also annoyed that my employees, due to their greater experience with the materials we used, could *cook* better than me. Maybe that was one of the reasons why I wanted to understand the basics, but I also think that is the task of a scientifically educated development manager, which I was or should be. It was not my job to be able to cook better than my *lab cooks,* who had been doing this for years. But we knew and understood no more about the physical and chemical basics of the production of mixtures of substances in polymer (i.e., *plastic*) systems than could be taken from technology monographs on polymers, and these also were ultimately just "cookbooks."

I quickly realized that we were dealing with a complicated area of *dispersion* with our products and processes, in other words: colloid chemistry and colloid physics. But none of the technology monographs, let alone the articles I could find, dealt with this. There was a lot of talk about *mixing,* about suitable machines, about process parameters and about achieved mechanical values, but nothing about what actually happened during these *mixing* processes at the microscopic or sub-microscopic level.

Even if it may seem strange at the moment: We will understand, through this only apparent detour, how chance and coincidence came into our world. You will gain many new insights from my research, as well as the outlooks far beyond that, which will be quite new to you. If there are also plastics and polymer experts among the readers here, only a few of them will have read about this before, they will be surprised and will maybe frown. But these new findings are also immediately applicable to many more products of our daily life, from mayonnaise to sun cream to hair shampoos etc. up to technical dispersions and emulsions of all kinds. And the understanding of such products will enable us to understand the origin of chance, coincidence and accidental events.

The secret is to be found here: All of these products consist of several *phases,* that is, of components that are not really compatible with each other, somewhat like oil and water; in other words, not like sugar dissolved in water, or alcohol diluted with water. The latter form a single homogeneous phase together with water. You cannot dilute gasoline with water, as it is not soluble in water. If you want to make good cocoa from real cocoa powder (that is, the powder that is nowadays almost exclusively sprinkled on tiramisu), powder that does not sink to the bottom of the cup after a few seconds, you have to proceed differently than if you were making sugar water. If you want to make mayonnaise or a beautiful sauce Béarnaise, you cannot simply pour everything that is on your shopping list into a pot and stir—the mayonnaise will be runny and unappetizing, the sauce will lump! Let's not even talk about a successful butter cream for a festive cake. Why? These are the questions addressed in dispersion research, colloid chemistry and physics. Only these products—as I have lamented in regard to my plastic compounds—have been developed without understanding the scientific basics. No problem for mayonnaise and sauce Béarnaise, but a serious deficit for a deeper understanding of dispersions in polymers, if one wishes to achieve it for development of products that would be more than trial and error (and I wished to).

You don't have to be a colloid chemist to make good cocoa or mayonnaise. I also learned how to make real cocoa from the beautiful cocoa powder long before I entered dispersion research scientifically and operated it industrially: We take this cocoa powder, add some sugar, mix the two dry powders together, take a tiny amount of warm water and prepare a paste from the cocoa-sugar mixture. The less water, the better, so you can disperse the cocoa with the help of the dissolved sugar! Then, we take the hot milk, pour only a very small amount into the cocoa paste at first and dilute it gradually. This will result in a delicious cocoa that your children and grandchildren and their friends will love, just like mine do. It tastes much better than if using the so-called *soluble cocoa powder*, which is also not soluble, but rather an industrially prepared powder that is easier to disperse, but also yields a thinner and much too sweet cocoa.

As I said, you don't have to be a colloid chemist to be able to do this and earn the recognition of the friends of your children and grandchildren. But to understand how chance and coincidence come into the world, why they dominate the entire world, at least a little deeper understanding about the principal difference between milk and mayonnaise on the one hand and sugar water on the other hand will help.

Because milk or mayonnaise or sauce Béarnaise on the one hand and sugar water or watery diluted alcohol on the other hand are completely different types of systems. As different as a living tree and a pile of useless ashes. From this follows much more, and this is what the next chapters are about. I will explain everything very simply and refer interested readers to short sections in the appendix.

Let's start with how "vinyl records" were mass-produced. Such records were the only music recording media for many decades. There was a drastic decrease in demand at the end of the 1970s due to the magnetic tape compact cassettes, and by the '90s, CDs had practically replaced both of them completely. After the phonograph record was invented, it achieved a market breakthrough when, around 1900, its production from a chemically modified natural resin, shellac, became possible. Even then, the company Zipperling was an important leading raw material supplier. In the 1950s, the process of replacing shellac with PVC was completed. Zipperling developed into the world's second largest supplier of vinyl record molding compounds based on polyvinyl chloride copolymers. These were not materials made from a single type of plastic, but mixtures of several polymer types, fillers and all kinds of more or less necessary additives. To be honest: I, as a chemist who had been awarded a doctorate only three years earlier, was very disturbed by the way in which we were producing them.

For almost every order, new raw materials were bought, each time from a different source, because suddenly some type from Czechoslovakia or from somewhere else was cheaper than the comparable raw material from France (Atochem) or southern Germany (Wacker-Chemie). Only, *comparable*, mind you, not *identical,* i.e., the proportions of the two or three PVC types and the proportion of the fillers had to be varied accordingly, on the one hand, to achieve approximately the same pressing behavior, and on the other hand, to reduce costs. A never-ending trial and error, to which the guardian of the quality of our record molding masses devoted himself with much love and enthusiasm, but completely erratically. Together with the purchasing and sales manager, who controlled the record molding componds department like a small principality of his own, he tinkered with recipes that, of course, only worked after several changes, which often required many days or even weeks. I didn't like this *fiddling.*

"How can you measure pressing behavior?", I asked him one day.

"You can't measure it, you can only see it when you look at the press cake before pressing and after pressing the record, whether all the grooves are cleanly pressed out" he replied.

I didn't believe that, and thought that pressing behavior must be a question of viscosity. So, I insisted on buying a rheometer—a quite complicated laboratory machine, expensive for our standards at the time. I learned a lot about rheology (the science of flow) and started with systematic measurements. Quickly, I began with a very large project in which I wanted to determine the influence of the individual components on the viscosity. Because I had *seen* something in the first measurements. I felt—that was all it was, it was just a feeling—that there must be a function, an algorithm with which I could predict a viscosity, and thus the pressing behavior.

When evaluating the data, at that time still manually with paper, pencil and calculator, I discovered a clear dependence of the resulting viscosity on those of the starting materials. The big surprise was that I could even assign a finite viscosity contribution to the non-melting fillers in spite of their therefore infinite viscosity (see Appendix 2, Part 2). I now wanted to convert the precise knowledge into a tool that would be easy to use for development and production, but how? And yes, there was chance again! We already had a computer, a huge, very slow machine that took up half a room, our *color computer.* With it, we could measure and compare colors, assess color differences. So, I was initiated into the use of this mysterious machine. I was able to save my data on a *Floppy Disc*[5] and to design a simple program in *BASIC*. With this, you could increase and decrease the proportions of any component in order to achieve a viscosity target value. This value described a record molding compound that could be pressed particularly well into records.

---

[5]The younger readers will certainly have to do some research to find out what that was.

Of course, no one in the company believed that my program could produce any useful results. Even the general manager and above all the commercial director of the record molding compounds department were extremely sceptical. So, I put it to a test with an order for 100 tons of record molding material that was to be loaded short-term in the Hamburg harbor for Brazil. I suggested a comparison between a formulation on an empirical basis, the usual starting point for subsequent fiddling, and an approach based on the results of my calculations.

The two formulations, i.e., the respective proportions of the necessary components, differed like day and night. And lo and behold, already in the evening, it was clear, measured with the rheometer and practically tested with the laboratory record press: My calculation result landed exactly on the target value, and the records were pressed out more perfectly than even I had expected. The proposal of our quality guardian, based on practical experience, landed miles away from the target. Instead of, as usual, trying to approach the target slowly over days with trial and error attempts, often not quite achieving it, and then starting a risky production because, for example, the departure of the ship in the Hamburg harbor for Brazil was imminent, it was now possible to perform a calculation that only took a few minutes. After that, one test in the laboratory was enough for confirmation, and production could start immediately. That was a real and also applied research! Only one year after my entry into the company and only eight months after the introduction of short-time work, we were able to hire a dozen new employees.

"Dispersion"—Practically Unexplored, but No Witchcraft!

For me, it became increasingly clear which knowledge gap I wanted to address with my research: What actually happens during the dispersion of substances, be they pigments, additives, fillers or polymers compatible with a matrix polymer? And above all: What happens at the interfaces that then form inside the generated "mixture"? I already had a certain idea of this, but it was still very nebulous. In Düsseldorf, I used vacuum during premixing and extrusion and examined the interfaces between the filler particles and the polymer matrix using a scanning electron microscope. I wanted to know back then whether the better properties were somehow reflected at the interfaces between the filler particles and the plastic matrix. I found some indications for this that became decisive elements in the patent application. Now, with the surprising discovery that even fillers contribute an *apparent viscosity* to the resulting overall viscosity, it seemed to me that this was also due to the interfaces—but I was still completely unclear as to why this could be the case.

I began to forge a plan of what I wanted to research and how. Already when I was still in Düsseldorf, my friend had sent me a copy of a publication from Paris about the discovery of the first electrically conductive polymer, polyacetylene, which had electrified me. The first polymer that conducts electricity by itself, without the addition of conductive fillers. Coincidences played a key role in this discovery as well, as was expressly stated in the press release of the Nobel Committee—when the three discoverers were awarded the Nobel Prize more than twenty years later.[6] A coffee break during a conference by chance brought the two US professors Alan MacDiarmid and Alan Heeger together with the Japanese Hideki Shirakawa for the first time, during which the Americans were made aware of polyacetylene (at that time, still not conductive). They had previously been researching an electrically conductive *inorganic* polymer, because they were interested in *one-dimensional electrical conductors*. In the subsequent work, iodine was *accidentally* used at 1000 times the concentration—contrary to the actual intention—and *by chance,* an employee took a conductivity measurement. Lo and behold, the first *organic* electrically conductive polymer was discovered!

I had been working, albeit initially in a very amateurish way, on the development of filler blends with polyethylene and polypropylene. While I researched more and more literature about the conductive polymer, an idea developed in a brain segment apparently outside my areas of conscious thinking: What would it be like to create a blend of polyacetylene with a conventional polymer such as polyethylene or polypropylene, just as with the inorganic fillers, only that the conductive polymer would then be the conductive component in a plastic blend? Occasionally this thought also flashed into my consciousness.

I read in the publications that polyacetylene, which was then the first and initially only electrically conductive polymer, was sensitive to oxygen, oxidized very easily, and was thus destroyed. In addition, it was not soluble and could not be melted. So, I started fantasising: "I'll just mix it under vacuum, like I did it with the fillers."[7] At some point, I went one step further in my thoughts: not *mixing,* but *dispersing,* even though I myself was not quite clear what that exactly meant—apart from the rather banal descriptions in books and articles, which always only dealt with *mixing* and *distributing.* This type of idea that occurred to me is called *analogy conclusion.*[8] Many people may think: "That's easy, just do it like with the fillers." If all analogies were easy, there would be (almost) no inventions. Because you have to come up with an analogy in the first place. Apart

---

[6] https://www.nobelprize.org/prizes/chemistry/2000/popular-information/

[7] I soon also used the technology of mixing under vacuum and extruding under vacuum with Zipperling and developed an electrically conductive, antistatic PVC, which was used in the electronics industry. For this purpose, we mixed PVC and so-called conductive carbon black (i.e., a carbon black with a very large specific surface area) and some other necessary components under vacuum, and also fed the finished mixture under vacuum to a twin-screw extruder. We produced hundreds of tons of this product over the course of a few years and had a monopoly. In the development on a laboratory scale, it became clear to me: Such a product could only be mixed and extruded under vacuum. So, we needed small (for the laboratory) and large (for the production plant) vacuum equipment as quickly as possible, including a so-called double funnel. The upper funnel serves as a lock; it is filled under normal pressure and also set under vacuum. A large valve opens, the mixture falls into the lower funnel, from where the extruder is fed—also under vacuum! So, the highly carbon-containing mixture cannot be oxidized, and the PVC does not decompose.

[8] https://en.wikipedia.org/wiki/Analogy

from me, no one has tried to implement such an idea—either that or they failed without making it public.

This still quite vague plan, which took root in me, also required me at the same time to lead the strategic reorientation of the company so it could become profitable again. So, on the one hand I had to make sure that we developed the new business area; on the other hand, I designed first concepts in my mind of what to research and how to research it. Because I really wanted to understand scientifically what we were practicing industrially. I quickly found out that there was no real basic knowledge in the field of *plastic compounding*[9], that is, the production of thermoplastic (heat-mouldable) raw materials, which contain various plastics, fillers, pigments and/or additives. These were, strictly speaking, dispersions that were viscous at processing temperature and solid at room temperature. I had discovered a large, almost completely unexplored field through a series of accidental events of the past years and just as coincidental ideas that had great practical importance, industrially and in everyday life.

I had been a patient tinkerer as a child and teenager. We were an eight-person family, and according to my naturally distorted memory, I was almost always the one who had to shine everyone's shoes. That was time-consuming and boring, so I *invented* a shoe polishing machine. I found an old washing machine motor somewhere, wrapped the flywheel in fur, screwed the whole thing onto a wooden stool, and had a self-made polishing machine. The tinkering and the constant repairs took ten times as much time as I saved by now polishing mechanically, but it was by no means as boring. Especially when the stool with the mounted and rotating motor began to wander around the kitchen while polishing.

In the youth center, where we met as scouts, I found an old, small telephone exchange in the basement. I took it home with me. It was a bluish-gray box that was about half a meter by almost a meter and about 25 cm deep. Inside were about forty relays. Each of the relays had twelve contact fingers that somehow had to be able to contact the neighbours so that, for example, you could call from extension 1 to 2. Only: It didn't work. Most of them were obviously bent. With endless patience and fiddling, I found out which fingers I had to bend a very little bit, correcting so that they all—that is, several hundred!—were the correct distance apart again and you could call from one extension to the other. Now, I laid copper cables throughout the house and connected the telephones that I had also found in the youth center basement. It was a lot of work, but it was crowned with success. The parents could now call from the 1st floor, for example, to the basement, where two of my brothers and I each had a room, and announce the meal—or remind us of any tasks to be done.

I felt that my research would require a similar patience. As for all situations in life, the Chinese for this have a *Cheng Yu* for this as well: 铁杵磨成针 (tiě zhù mó chéng

---

[9] https://en.wikipedia.org/wiki/Plastic_compounding, which shows a purely technical (engineering) point of view of "compounding", with no reference to any basic understanding of "dispersion".

zhēn). This means something like "grinding an iron rod into a needle". The following story[10] is based on it:

> During the Tang Dynasty (唐朝 Tángcháo), there lived a famous poet Lǐ Baí (李白, 701–762), who, as a child liked, to play and did not concentrate on books. One day, having run away, as he had done before, he saw an old woman grinding an iron rod on a large stone. Curiously, he asked: "Grandma, what are you doing?" Without taking her eyes off her work, the old woman replied: "I want to grind this iron rod into a needle to sew with." Unsatisfied with this answer, Lǐ Baí asked further: "What? You want to grind this rod into a needle?"— "If I grind today and tomorrow again, it will get thinner and thinner. Then, I will finally reach my goal one day."
>
> Lǐ Baí realized the value of this subtle message and decided from then on to learn diligently. In the end, he became a great poet.

My boss supported the project, even though he had no expectations that I would be able to discover anything valuable at all. But he had noticed that I had made important contributions to a appreciable improvement in productivity. Even more, he knew that I had saved Zipperling from bankruptcy twice within a few months and then started the company's first big, very lucrative business with the big chemical company following the new strategic direction, with which we had pulled ourselves out of another bankruptcy quagmire by our own hair.

The company grew. I hired a chemist, and I took on the first high school graduates as trainee chemical laboratory assistants. The very first one would go on to work for me for over thirty years. Slowly, my development laboratory evolved into a larger and more efficient department, and I began referring to a small part of it as a "*research* group". We started with the synthesis of polyacetylene.

I had already drawn up and submitted an application for research funding from the Federal Ministry of Research. Research on conductive polymers was a hot topic worldwide at the time, as are *graphene* and *high-temperature superconductors* today. My boss was sceptical, but promised: If I managed to raise funding against all expectations, he would take me out for a meal in Hamburg's oldest oyster restaurant. I didn't know anything about oysters at all; I didn't even know if I should be disgusted by them or at least try them. I accepted the invitation, which Mr. Petersen believed he would never have to honor. But a few months later (after many tough discussions in Bonn), I had the funding approval, and we slurped oysters and champagne together with our finance director, who was allowed to celebrate with us. After two dozen of these delicacies, there was, "as dessert", some schnitzel with fried potatoes. Petersen: "The most important thing here is not the schnitzel, but the fried potatoes!" Since that night, I am a fan of oysters.

---

[10]Taken from the collection of proverbs of the East Asia Institute, see https://bit.ly/3sD3KvW, reprinted with kind permission of the institute.

Dispersion—one after the other

From today's perspective, my idea of dispersing conductive polymers appears as a simple matter. Namely, because I have shown that it works, and because it is generally known that other substances, e.g., pigments, are routinely dispersed. Unfortunately, most people think about dispersion something like this: "It's done by stirring, and it's really just mixing." Even if, in industry and in the daily kitchen of almost every household, a great many things are dispersed all of the time, when I started dealing with it, the scientific understanding of *dispersion* was not much further advanced than it had been 100 years earlier. And that meant: Regardless of where one dispersed in industry or research, one was not much more further advanced than in the domestic kitchen, where one could produce the desired result quite neatly with a lot of experience and instinct. But I was not satisfied with that. So, if I told someone that I wanted to disperse conductive polymers, my conversation partners shrugged, at best: "Yes, well, you can try it, it is quick to test, but what's the point?"

First of all, let's look at the ideas that prevailed at the time when I wanted to implement my dispersion hypothesis. The fact that I am quoting here from the Nobel Prize justification from the year 2000—about twenty years after my first dispersion ideas in 1978—shows the tenacity of original concepts (more on this in Chap. 5). You may believe me that not only were these the dominant concepts around year 2000, and earlier in the 1980s and 1990s, and essentially representing what was considered to be correct by practically all researchers long since after the discovery after these new polymers, but even now, more than 40 years later, with the circle of researchers having become much smaller, these ideas continue to haunt us:

"For a polymer to be able to conduct electric current it must consist alternately of single and double bonds between the carbon atoms. It must also be "doped", which means that electrons are removed (through oxidation) .... These "holes" or ... electrons can move along the molecule – it becomes electrically conductive."[11]

---

[11] https://www.nobelprize.org/prizes/chemistry/2000/press-release/ In this representation, several aspects are not correct, and this has nothing to do with the fact that it has been simplified here for the press and the readers. For one thing, there are (and have been for a long time before the Nobel Prize was awarded, e.g., by my company Ormecon) polyanilines, in whose structure nitrogen atoms and benzene rings alternate rather than single and double bonds. Secondly, "doping" is a false term, because oxidation in the case of polyacetylene is a real chemical reaction, while "doping" in semiconductors refers to the addition of a trace of impure foreign metal. Thirdly, for some types of polymers, like my polyaniline, conductivity is not achieved by oxidation, but by protonation (= by adding H^+ in form of strong acids). Furthermore, it was already known at that time—but was widely ignored, and I could also show this in regard to my polyaniline—that the electrons do not move "along the chain", but three-dimensionally through space, which is by no means a linear stretched chain, but a nano-sphere with about 10 nanometers in diameter, in which the chains (at least, for my polyaniline, for which I could prove it) have short helical screw structures, see Appendix 12, Part 1.

Since research had its origins in the search for *one-dimensional conductors*, it was hoped and assumed that polyacetylene would have a long and straight linear chain structure. Based on transmission electron microscopy, it was thought to have a fibrillar, i.e., fibrous morphology. The chains could be arranged nicely in the fibrils—thus the assumption that had crystallized into certainty, which had its origins in a hope—lengthwise and parallel, the ideal one-dimensional conductor was supposedly found.[12] The graphic reproduced here from the justification of the Nobel Prize [sketches (c) and (d) in comparison to diamond and graphite in (a) and (b)] was supposed to show this.

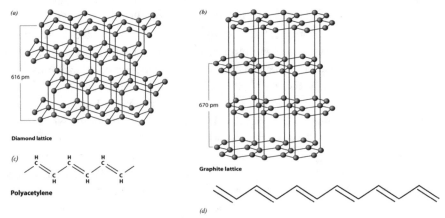

FIGURE 3
Three-, two- and one-dimensional carbon materials: diamond (a) and graphite (b) crystal lattices, and polyacetylene chain (c). An alternative way of writing polyacetylene is also shown (d).

This was identical to the assumption that the electron flow runs along the chain. The Nobel Committee published an animated representation of this (the following is a non-animated screenshot; in the link in the footnote, electrons shown in red jump along the chain).[13] In the text, it says that the electrons move along the chain.

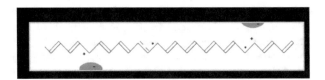

[12] Fig. 3 has been taken from the scientific background material that the Nobel Committee published: https://www.nobelprize.org/uploads/2018/06/advanced-chemistryprize2000-5.pdf, shown here with kind permission of the Royal Academy of Sciences "©Typoform/The Royal Swedish Academy of Sciences".

[13] https://www.nobelprize.org/prizes/chemistry/2000/popular-information/, with kind permission of the Royal Academy of Sciences "©Typoform/The Royal Swedish Academy of Sciences".

According to this idea, in the case of a *dispersion,* the fibrils and perhaps even the chains would tear, and then the conductivity would at least be reduced, if not completely disappear.

I pondered: If I wanted to disperse successfully, it had to be done as with the incorporation of so-called *conductive carbon black,* in order to produce antistatic plastics throughout the volume and not just on the surface of the products. There was an undisputed theory in industry and science (the percolation theory) according to which the carbon black particles were optimally evenly distributed throughout the plastic volume. Since the carbon black particles were considered "highly structured" —they had, so to speak, many "arms" —the carbon black should be dispersed in the thermoplastic material under shearing, but not too strongly, so that the particles are not destroyed.

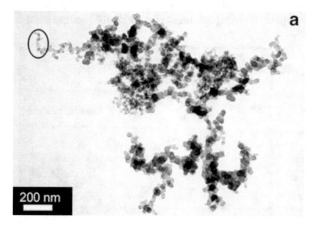

The transmission electron micrograph shown here makes it clear as to what the representatives of this theory base their assertions upon.[14]

[14] M. Spahr, R. Gilardi, D. Bonacchi, "Carbon Black for Electrically Conductive Applications", Chapter 19, Fig. 2a (p. 380) in R. Rothon (ed.), "Fillers for Polymer Applications", Polymers and Polymer Composites, a Reference Series, Springer Intl. Publishing Switzerland 2017; Reprinted with kind permission from Springer Nature. The year of publication shows that the idea sketched above is still widespread and predominant today. But the photo shown here does not describe carbon black as it is incorporated in the polymer (or plastic) matrix, but after the matrix has been dissolved and removed and the residue has been applied to a carrier, that is, many laboratory process steps later, i.e., we do not see the carbon black "as is" within the polymer matrix. See also the representation in a current brochure of the leading conductive carbon black manufacturer Cabot: https://www.cabotcorp.com/~/media/files/brochures/specialty-carbon-blacks/brochure-specialty-carbon-blacks-for-conductive-and-esd-applications.pdf

The theory used to explain the results[15] therefore works with *probabilities* with which such highly structured particles could meet. However, images like the one shown above and other comparable SEM photos were always taken of samples that did not show the carbon black as it was present in the incorporated state, but after the matrix had been removed with solvents and the carbon black had been placed on a substrate for examination. Of course, this only shows us what these particles look like *after* they have gone through another process history, and not how they actually exist in the manufactured plastic blend. Nevertheless, such photos were considered evidence of the high structure of the particles in the mixture. Then, it is said, the conductivity could increase by several orders of magnitude at the *percolation point*. The graph below[16] shows an exemplary course of the resistance (i.e., the reciprocal of the conductivity) and the widely held understanding of how it comes about: The structured and randomly distributed particles are supposed to eventually meet at some point, above which the composition would suddenly become conductive.

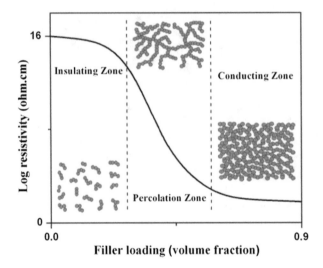

If I now wanted to disperse fibrils that were not structured, how should these meet, how should the electrons then pass from fibril to fibril, especially if the fibrils were torn apart due to the dispersion?

---

[15] https://www.researchgate.net/publication/258682829_Gelation_Mechanisms; it is significant that the percolation theory is considered a model for "gelation" here; I will show that this is not the right approach. For an overview about percolation theory, cf https://www.semanticscholar.org/paper/Introduction-To-Percolation-Theory-Stauffer-Aharony/66c431c2dfe70556dc021cf0e1b8a62 c2531922f.

[16] Fig. 10, p. 169 in M. Rahaman, A. Aldalbahi, L. Nayak, R. Giri, "Electrical Conductivity of Polymer-Carbon Composites: Effects of Different Factors" in: M. Rahaman et al. (eds), Carbon-Containing Polymer Composites, Springer Series on Polymer and Composite Materials, Springer Nature Singapore 2019, reprinted with kind permission from Springer Nature.

My idea was obviously quite naive and also contradicted all generally accepted scientific opinions. I must confess, to my shame, that I didn't think much about it, but above all, I wasn't worried. Because, first of all, I wanted to find out what polyacetylene looked like exactly.

To my surprise, I saw in the scanning electron microscope that the polyacetylene powder I had produced consisted of small spheres, really small, namely, only 100 nanometers in size. My powder contained no fibrils. Now, a suspicion began to dawn on me. I asked the Max Planck Institute for Polymer Research in Mainz, which was also researching in this area, to produce this substance for me exactly according to the usual method, which should lead to fibrils (see Appendix 5). The researcher there cut the resulting film and sent me half of it. We had agreed that he would conduct a *transmission* electron microscopic (TEM) examination there on site with his half, while I would look at my half with the *scanning* electron microscope (SEM). Now, the surprise was complete: In the Mainz TEM (i.e., in the *transmission*), a fibrillar structure was to be seen (picture on the left). With me, in the SEM (in *scanning* mode) a spherical structure was revealed (picture on the right): The fibrils therefore consisted of small spheres closely nestled together, strung together like beads on a string.

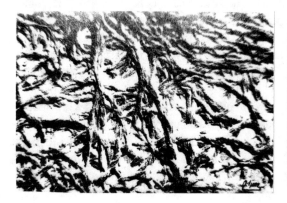 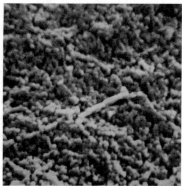

I published this finding, which I considered to be sensational.[17] It had to raise doubts everywhere about the assumed conductivity mechanism "electrons flow along the chain", because, if the smallest structures are not fibrils, but even smaller spheres, how are the chains arranged in the nano-spheres? And how do the electrons get from one sphere to

---

[17] https://www.researchgate.net/publication/230304769_Beitrag_zur_Diskussion_um_die_Morphologie_von_polyacetylen, Original source: Makromol. Chem. 185, 1265 (1984); an examination of the films in the TEM therefore tempts one to regard "fibrils" as morphological structural elements, because a shadow projection conceals the boundaries of the spheres.

another?[18] I experienced my first disappointment: Nobody, really nobody was interested in the spheres, nobody began to doubt the fibrillar structure and the conductivity mechanism. (The publication cited above was available 16 years before the Noble Prize honoring the discovery of conductive polymers. And after this publication, several additional ones followed, including oral presentations in conferences, showing ever more details about the spherical or globular structure of the primary particles in any and all conductive polymers.)

Despite the disappointment about the non-perception, this discovery inspired me to finally start with dispersion. I started playing with all kinds of organic solvents. It was known that polyacetylene was not soluble. Now I tested whether it could possibly become *dispersed* in any unusual solvents. I saw a few, not well reproducible positive indications, until, once again, an accidental event helped: A synthesis batch had not been dried properly, resulting in a wonderful, richly blue-colored transparent liquid. You could visually quite take it for a real solution, but a particle size measurement showed me about 100 nanometers, same as what I had seen in SEM—so I had achieved my first dispersion! Only much later did I realize why conductive polymers are so extremely difficult to disperse: The easier a substance is to disperse, the better its surface tension fits the surface tension of the medium in which the substance is to be dispersed. The smaller the distance, the better. Conductive polymers, however, as I found out many years later through complicated measurements, have a surface tension that is about ten times as large as that of water. A difference that perhaps helps you to understand the enormous difficulty of dispersing this class of substances. And by the way, it explains why they are *insoluble in principle*.

This principle—do not completely dry the polymerisation batch—, and a few process elements such as shearing and ultrasound and a minimum value for the polarity of the dispersion medium were of use to me as characteristic elements in my first basic patent on the dispersion of conductive polymers.[19] In it, I did not limit myself to polyacetylene, but, based on corresponding experimental data, I was able to extend the scope to all types of conductive polymers.

---

[18] Later, with even better resolution in the SEM, I found out that the smallest building blocks were not spherical nanoparticles 100 nanometers in size, but 10 times smaller again: 10 nanometers.

[19] German priority June 1984; European Patent: German priority June 1984; European Patent: https://patents.google.com/patent/EP0329768A1/de, US Patent:

https://patents.google.com/patent/US4935164A/en. With this patent, I was later able to sell licenses to large chemical companies, including Bayer AG, DuPont and a few others, even though, e. g., Bayer AG had developed a conductive polymer that I had not specifically mentioned, but which nevertheless was covered by the laims of my patent due to its chemical characteristics; and DuPont had started to produce some polymer blend using neutral polyaniline, which as well was covered by my patent.

My observations of the dispersion behavior in different solvents and in connection with my practical experience with antistatic carbon black compounds led me to first ideas: Dispersion has something to do with active, physical interactions at interfaces and with the *compatibility* of interfaces (see Appendix 6).[20]

The fact that I had succeeded in dispersing conductive polymers—it was a world premiere! —, even if only in low concentration, did not elicit any reaction in the worldwide research community, which, by that time, comprised thousands of researchers. It was all uninteresting to them, not even worth ignoring. I actually understood that, because I would first have to show that I could also produce macroscopically conductive samples in this way. So far, I was not there yet. I was optimistic that I could do it, but I was not fully aware of the difficulties I would have to overcome.[21]

The fact that one could also produce conductive plastic materials to a certain extent by dispersing carbon black in all kinds of plastics, was the source of all my optimism: by *dispersion* of carbon black, not just mixing it with plastics, that's a huge difference! But, at the same time, I knew that I first had to understand why one could achieve a certain conductivity with carbon black at relatively low concentrations.

The Discovery: Dispersions are Structured

All of what was read in the usual publications and that resulted from theory contradicted my experimental laboratory results and practical observations in production. I found out that an extremely good dispersion allows for extremely low critical concentration values, above which the conductivity increases abruptly. Quite contrary to theory, this critical concentration was reproducibly dependent on the type of plastic matrix: the more polar the type, the higher the necessary proportion, the less polar, the lower. Polyethylene showed the lowest, polyamide and polyacrylate the highest. According to percolation theory, it should not matter at all in which plastic the carbon black was *distributed*, it

---

[20] There are also a few bits of more detailed information about the breakthrough of my dispersion research a few years later, when I realized that a very special, extremely finely controlled polymerization process is required to reliably and reproducibly obtain dispersible conductive polymers.

[21] One of the biggest problems was the reproducibility of the synthesis of the conductive polymer polyaniline. Eventually, I noticed that it became more reproducible when we cooled it. This led to an elaborate multi-year project in which we examined the synthesis conditions in detail. A patent (https://patents.google.com/patent/EP0329768A1/en) and, in the course of the long-term work, a computer-controlled synthesis, which is still used commercially today, emerged. I jump ahead to the next chapter: The synthesis is a classical non-equilibrium process. The synthesis methods used by the worldwide research groups were practically equilibrium processes. They lead to non-dispersible conductive polymers. A few bit more information can be found in Part 2 of Appendix 6.

would always result in approximately the same concentration of the breakpoint to conductivity. Such a result is predicted by this theory because it only provides for an *optimal distribution* and no interactions whatsoever between the *distributed* substances and the environment. But these interactions obviously take place, because the polarity of the environment influences the value of the critical conductivity breakthrough concentration, i.e., the point at which the carbon black particles meet. There must be some kind of interaction, but which one? The particles are not just *distributed,* but *dispersed*! This was shown dramatically in my further investigations.

To even my own surprise, I initiated the density profile to be measured with increasing carbon black content. It should increase linearly, as one would expect. I no longer remember how I came up with the accidental idea to carry out such a simple—seemingly pointless?—measurement series, because I did not expect anything exciting! Probably, I just wanted to collect as many data as possible, because measurement results are, after all, the basis of research. My laboratory assistant asked: "Why should I measure that?" I replied: "I do not know, just do it, it will not hurt."

And lo and behold, I discovered an incredibly surprising phenomenon: The density did not behave linearly along the linearly increasing concentration, but showed a kink in the course.[22] Exactly at the critical concentration, at which the conductivity increases abruptly, the increase in density stagnates, with the density even decreasing slightly in some cases; above it, it increases again linearly, but with a lower slope than below the critical concentration (cf. an original plot in Appendix 7, Part 1). Why should this only apply to finely dispersed dispersions of carbon black? This question prompted us to also examine a concentration series of a pigment dispersion in polystyrene. It was not quite as surprising: This density/pigment concentration plot also showed a kink. Of course, this pigment/polymer matrix dispersion does not have a conductivity increase at all, but apparently shows a similar behavior that led to conductivity with carbon black in a polymer. I had also invented a two-phase system, consisting of two mutually compatible polymers. In this there, two conductivity breakthrough concentrations could be found. It was now logical that it accordingly showed two density kinks (cf. 2nd graph below).

---

[22] B. Wessling, "Further Experimental Evidence for a Phase Transition at the Critical Volume Concentration", Polym. Eng. Sci. Vol. 31, No. 16, (1991) pp. 1200–1206, full text available at https://www.researchgate.net/publication/243341314_Electrical_conductivity_in_heterogenous_polymer_systems_IV_1_a_new_dynamic_interfacial_percolation_model.

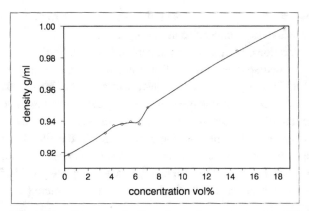

Fig. 4. Density-concentration dependance in a pigment 13/PE system.

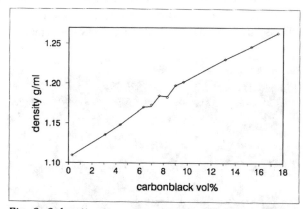

Fig. 2. 2 density stagnation points in the blend of Fig. 1.

We made another series of measurements: If we pyrolysed the concentration series, at low concentrations, more pyrolysis residue remained than we had used as conductive particles amount to be dispersed; the higher the carbon black or polyaniline concentration, the lower the additional residue per dispersed component content. All of this was clearly in contradiction to percolation theory, and it was also completely incomprehensible—a huge puzzle had built up in front of me. It took me almost a year to solve it.

First, I decided to look *into* the finished solid dispersions to see what kind of structures the carbon black particles had actually built in there. I didn't want to just *assume* anymore, I wanted to *see* and *know*! I looked at the products with a scanning electron microscope after breaking them under liquid nitrogen, so that I could examine internal fracture surfaces. And I did this with samples that contained different concentrations of carbon black: below, directly at, and above the critical concentrations.[23] I also looked at the first samples with a low concentration of polyacetylene.

At low concentrations, both substances, carbon black and polyacetylene, are present in the form of isolated spheres about 100 nanometers in size or smaller (see photos in Appendix 7). What? I thought the carbon black particles were highly structured and had lots of "little arms"? Then, it struck me that I saw a great deal many more particles in the samples, on the fracture surfaces, many more particles than the low concentration—only 0.5%—would suggest. This could only mean one thing: The carbon black dispersed into individual spheres or the polyacetylene also present in very small isolated spheres was concentrating in certain areas, so there had to be areas completely void of such particles! Where was the "homogeneous distribution" as claimed everywhere (and required by percolation theory!)?

But what was particularly surprising was that the loos (at low concentration) distribution of dispersed nanoscopic spherical particles in compositions with a concentration directly at or just above the critical volume concentration showed highly complex structures! There, I saw *strings of pearls* that branch off, and these again in a much higher concentration than just the 6% we had worked into it. How could this be?

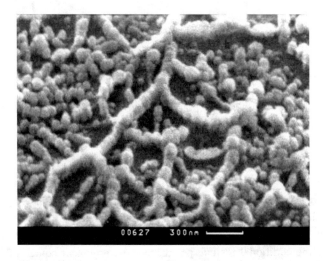

In Appendix 7, I describe in a little more detail how, for months, I tried to solve this puzzle. It was a tormenting time for me, because I felt that I was close to finding something new and very important, but I couldn't find it. Again and again, I literally had stomach pains, so strenuously and sometimes desperately did I work mentally on the clarification of the question. I developed mental image after mental image, nothing fit together… Finally, I solved the complex puzzle during a twelve-hour wait at Newark Airport (New

---

[23] Fig. 5, 6 from B. Wessling, preceding footnote.

Jersey, USA) due to a snow disaster (see Appendix 7). While it was incredibly chaotic at the airport because all of the passengers, like me, had already been delayed for days, with only a few getting a flight and many becoming hysterical, I was calm as a cucumber. For hours, I sat on the floor leaning at the wall, didn't move, didn't care about whether I got a flight or not; I hardly noticed the constant loud and meaningless announcements through the loudspeakers, nor the yelling and the hectic back and forth of the thousands of passengers around me, but I scribbled sheet after sheet with new drawings, crumpling up 99% of them and throwing them into an already overflowing wastebasket. One sheet remained (sketch diagrams in Appendix 7):

Below the critical concentration, the carbon black and the conductive polymer nanoparticles are completely dispersed as nanospheres. This means that a very thin, very firmly adsorbed shell of molecules of the matrix polymer has formed around the carbon black or conductive polymer particles. The conclusion that it is a firmly adsorbed shell results from the pyrolysis measurements. The dispersed particles are closely packed together in layers, which—as someone from the Ruhr region—I called "seams" (i. e., like "coal seams"). At the critical concentration, the shells of adjacent particles unite, a long tube is formed, the spheres bump against each other, and electrons can now jump from sphere to sphere. These chains look, as I often described it to make it more understandable, like a snake that has swallowed several golf balls. This creates cavities that cause the kink in the density curve. Gradually, it became clear to me that it must be a non-equilibrium system.

In this picture, you cannot see "seams" like the coal seams in the Ruhr region, but rather the result of deposits that took place over millions of years during the cretacious period and also formed layers. Later, when the Himalayas were formed, these layers were uplifted by the same shifts of continental plates. After that, they underwent changes due to erosion and oxidation, so that these beautiful landscape forms and colors were created. These are also results of non-equilibrium processes that continue to take place. Equilibrium processes cannot create something like this. (Photo by the author in the Danxia Shan National Park, China, a UNESCO Global Geopark)

The equilibrium concept claims: After dispersion, all particles are statistically evenly distributed throughout the volume. But I had discovered: The smaller the particles are, that is, the better one disperses, the lower is the critical concentration at which conductivity occurs. We were able to achieve values down to only three percent, later, with our polyaniline, even as low as 1.5%. But, according to the percolation theory, well-dispersed nano-beads would require far more than fifty percent concentration in order to allow conductivity, and, according to this theory, irrespective of particle size. But the critical volume concentration is, in fact, extremely dependent upon the type of dispersion matrix, that is, the dispersion environment and the dispersion degree. Therefore, a polar environment causes a higher jump concentration because the beads are better wetted; the tube with the particles touching each other only forms at higher concentrations because the shell adheres better.

These facts, which have been 100 % experimentally confirmed, cannot be explained by the percolation theory, a theory for describing equilibrium states.

This finding now also explains why non-melting components in a polymer mixture contribute to apparent viscosity despite their own infinitely high viscosity (because they do not melt), as I had discovered very early on with Zipperling's record molding compounds. The explanation can be found in Appendix 7, Part 2.

It should now be clear why I took this long excursion into the chemistry and physics of dispersion. It is extremely important for our understanding of the world to see that the previous idea of *equilibrium* mostly does not describe reality. Dispersions and emulsions are examples of many other non-equilibrium systems that are still mostly misunderstood as equilibrium systems.

Why is this so crucial? *In non-equilibrium systems, completely different types of laws prevail than in equilibrium systems. And in these other behaviors, we will find the causes for the emergence of chance, coincidence and accidental events.* An important characteristic of non-equilibrium systems is the presence of complex and usually highly interesting structures.

Thus, the concept of dispersions differs considerably when one looks at the distribution of the dispersed particles in "equilibrium" and in "non-equilibrium" (it is irrelevant how small they really are—if we are dealing with 100 nanometer small particles, the axes are correspondingly small, but the picture looks exactly the same!):

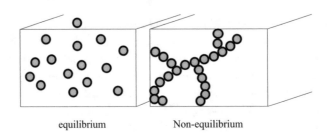

equilibrium   Non-equilibrium

Mayonnaise, Béarnaise Sauce, Yogurt and Cheese

We clearly see the fundamental difference in the distribution of substances in equilibrium and in non-equilibrium. For example, an equilibrium distribution is present in a *solution* of sugar *molecules* in water. Here, it is not particles, but individual, truly dissolved molecules, which are surrounded by water *molecules*; these are much smaller than 1 nanometer. They are simply homogeneously statistically evenly distributed. Non-equilibrium systems look quite different, as, for example, the arrangement of oil droplets in mayonnaise shows. There are two crucial characteristics that mark the system *mayonnaise* as a non-equilibrium system: The order of addition of the components is mandatory! You cannot put the olive oil first, add the juice of a squeezed lemon, stir, and then work in the egg yolk, or even put everything in a bowl at once and stir—no, first the egg yolk and then the lemon juice, then add the oil dropwise, not in one shot! Furthermore, you *have to* expend a relatively high amount of energy, light stirring is not enough, you have to really beat it with a whisk, because you have to break up the oil droplets, the finer, the better for the consistency of the mayonnaise. With a sugar solution, on the other hand, it does not matter at all how you proceed. You can first fill the cup with water and then add a few grains of sugar, or vice versa. And you don't even have to stir, waiting is enough. It is only your impatience that makes you want to stir, after all. Light stirring is quite enough if you are impatient; you don't need a whisk or a food mixer to make a sugar solution.

Mayonnaise is an "oil-in-water emulsion", similar to a sauce Béarnaise, for which you first add warm, liquid butter dropwise to an egg yolk/water or /white wine mixture and beat the resulting emulsion vigorously again. You disperse!

A few years ago, by chance, I read an article in a physical journal, entitled *The Mayonnaise Effect.*[24] In my opinion, the article's fundamental error was the attempt to explain the properties of an *emulsion* using graphics, formulas, and arguments from the description of *solutions*. Therefore, it was not surprising that the equations applied failed for the exact condition that was supposed to be described according to the article's title. The author attributed this failure to the *Mayonnaise Effect*, or the *jamming* (i.e., the "becoming-like-jam" process). I wrote to the editor that I was of the opposite opinion, briefly outlined my position and was invited to write a more detailed comment. This was accepted and published very quickly.[25] I explained that it is not possible to describe a non-equilibrium system like mayonnaise, namely, an emulsion, using formulas for an equilibrium system, in this case, a salt solution. Nor could a supersaturated solution be taken as a model for an emulsion, and the failure of an equation does not explain the sud-

---

[24] K. Wynne, "The Mayonnaise Effect", J. Phys. Chem. Lett. 2017, 8, 6189–6192, download is possible here: http://eprints.gla.ac.uk/153368/7/153368.pdf.

[25] B. Wessling, "Comment on 'The Mayonnaise Effect'", J. Phys. Chem B 2018, 122, 10, 2821–2823, here is the full text: https://pubs.acs.org/doi/10.1021/acs.jpcb.8b01006#

den increase in viscosity, but only that the equation cannot be applied here. The viscosity behavior of emulsions, I explained, can be predicted according to my equation developed thirty years earlier for the calculation of the viscosity of multi-phase systems (dispersions, emulsions):

$$\textbf{\textit{ln}}\ \eta_{(res)} = (1\text{-}x_n)\textbf{\textit{*ln}}\ \eta_{(matrix)} + x_n\textbf{\textit{*a}}_n \qquad (1)$$

where η is the viscosity resulting$_{(res)}$ from the matrix$_{(matrix)}$ (continuous phase) viscosity plus the contribution **a**, of component n which has the volume concentration x_n, while a_n, the viscosity contribution of component n also has the form of the natural logarithm of a viscosity, but this is only an apparent viscosity because also solids exhibit a similar contribution factor a, and for liquids n (in emulsions) a is not the viscosity of the pure component n.

I showed a microscopic image[26] (generated with so-called confocal microscopy[27]) of mayonnaise, into which I had schematically marked two-dimensional parts of the three-dimensional arrangement of the oil droplets with a red line.

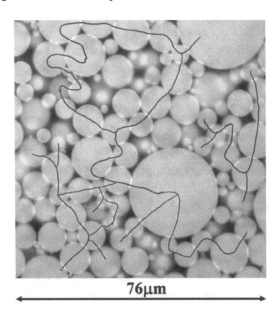

76μm

Wherever fine light areas can be seen between the droplets, something very similar has happened to what I had discovered with my carbon black and conductive polymer dispersions: Instead of being isolated from each other, they have—because they have

---

[26] http://www.ianhopkinson.org.uk/2010/05/understanding-mayonnaise/ Reproduced with permission from the American Chemical Society.

[27] https://en.wikipedia.org/wiki/Confocal_microscopy

"become too numerous" above a critical concentration—partially come into very close contact with each other without completely merging! Thus, they form a three-dimensional network structure, and this makes the mayonnaise stiff. Fortunately, the author I had criticized was then able to publish a very short answer to my reply in the same issue of the journal[28], in which he explained that my differentiation of solutions and dispersions as equilibrium or non-equilibrium systems seemed correct to him.

We can safely assume today that all natural, as well as all artificial, dispersions and emulsions are non-equilibrium systems with a complex structure and show a completely different behavior than solutions (equilibrium). This can also be seen in the completely different flow behavior: Pure liquids and solutions show a viscosity that is referred to as "Newtonian"[29], dispersions and emulsions behave quite differently and are referred to as "non-Newtonian" fluids[30]. One of the most commonly known is ketchup: If you open a full bottle and turn it upside down, the ketchup remains unimpressed in the bottle; if you shake it (that is, exert a force on the ketchup), it flows out. Ketchup is a thixotropic fluid. Starch dispersed in water behaves quite differently—starch is insoluble in water, but can be dispersed in it! At a concentration of about 3:2 (ratio of starch to water), you get a mush. If you run a fork through this mush very quickly, it thickens! Now, if you pick up the lump on your fork, hold it up and let the unruly mixture calm down, it will begin to drip off the fork after a while. Such a non-Newtonian fluid shows a *dilatant* viscosity behavior. Even with my understanding that dispersions and emulsions are highly structured, it is still puzzling why some of these systems are dilatant, others thixotropic. Yes, with additional shear, more network structures are formed, or existing ones are (partially) destroyed, but why is it this way with one and that way with the other?

Now, let's look at other interesting emulsions under the electron microscope: for example, freeze-dried yogurt,[31, 32] (more examples in[33] and[34]). The orange-colored particles are the bacteria that made the yogurt for us, everything else there the solid components

[28] K. Wynne, "Reply to 'Comment on 'The Mayonnaise Effect'''", J. Phys. Chem B 2018, 122, 10, 2824, fully readable here: https://pubs.acs.org/doi/10.1021/acs.jpcb.8b01428

[29] https://en.wikipedia.org/wiki/Newtonian_fluid

[30] https://en.wikipedia.org/wiki/Non-Newtonian_fluid

[31] http://www.magma.ca/~scimat/science/Yogurt.htm; there are also more pictures here, including of milk itself; reprinted with kind permission from Dr. Miloslav Kalab.

[32] https://www.youtube.com/watch?v=WhXY3YbY-08, reprinted with kind permission from Dr. Miloslav Kalab.

[33] https://www.researchgate.net/publication/333739281_Milk_from_different_species_on_physicochemical_and_microstructural_yoghurt_properties and also here: https://www.researchgate.net/publication/307904322_Measurement_of_Casein_Micelle_Size_in_Raw_Dairy_Cattle_Milk_by_Dynamic_Light_Scattering/figures?lo=1

[34] https://imagenavi.jp/search/detail.asp?id=70090105

of the milk: They have formed a three-dimensional network structure that causes the typical viscosity of the yogurt.

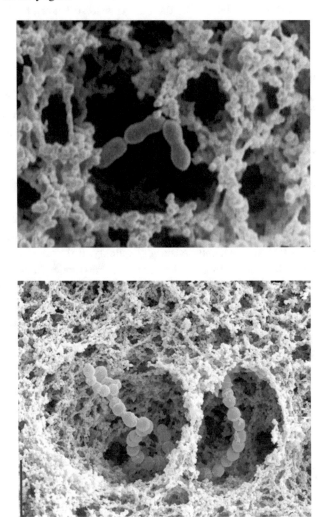

The difference between milk and mozzarella, more precisely said: the process of milk becoming mozzarella, which corresponds to the transition from a liquid to a relatively solid emulsion, can be illustrated as shown below.[35] In milk, the casein protein droplets

---

[35] https://www.slideshare.net/nsutaria/the-food-science-behind-mozzarella-cheese, see also http://stsimondicroce.blogspot.com/2015/10/di-croce-gr.html (Science 1.4), reproduced with kind permission from slidedoc.com according to their terms of use.

("micelles") are isolated from each other. In cheese, they then form pearl chains, and finally three-dimensional networks:

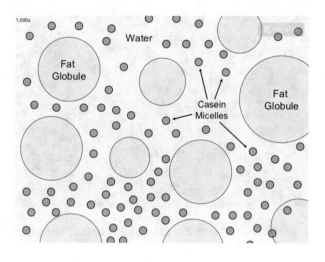

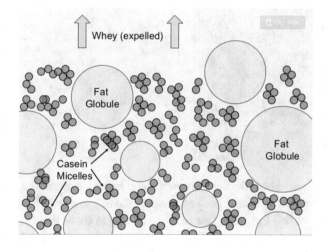

Electron microscopic photos of the casein network in cheese[36] and tofu[37] show comparable structures.

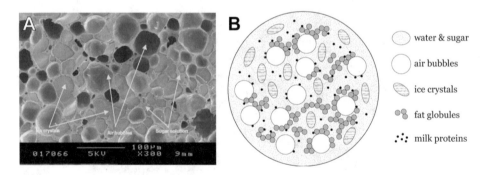

Finally, I don't want to hide the fact that even much-loved ice cream is a non-equilibrium system with a complex structure, as can be seen in the image above.[38]

Everywhere, we see structures that are very similar to those that I found in my plastic dispersions. Of course, they are different dispersed or emulsified substances in a different environment. But we can clearly see: The particles or droplets are by no means uniformly, statistically homogeneously distributed, but are forming complex three-

---

[36] http://www.magma.ca/~pavel/science/Foods&bact.htm, printed with kind permission from Dr. Miloslav Kalab.

[37] https://www.thevetgroup.com.au/more-cheese-gromit/casein-matrix/;   casein   https://bit.ly/3pQlYGn

[38] https://scienceandfooducla.files.wordpress.com/2013/02/icecreamstructurefull-02.png   printed with kind permission from Dr. Liz Roth-Johnson.

dimensional networks with different densities, different lengths of *pearl chains,* which then cause the different properties in flow behavior. The structures I had found corresponded to those that Ilya Prigogine had called *dissipative structures.*[39] I had discovered that polymer blends and compounds and all dispersions and emulsions, due to their non-equilibrium character, display such *self-organized* structures and are, in principle, chaotic-dynamic systems that can change their properties non-linearly after only apparently minor changes of one or more parameters, but do not necessarily do so. A non-equilibrium thermodynamic theory of dispersion did not exist at that time. It took several years of intensive research before I could submit first basic work. More about this later in Chap. 5, as well as details in Appendix 11.

Sparkling Ideas and Flashes of Inspiration

I have tried to avoid the impression that I had a well-thought-out and long-term research plan that I *simply* systematically pursued and worked through. Because that's not how it was. Although the microscopic and electron microscopic images of other dispersions did show that what I had discovered about plastic dispersions is apparently quite normal. Yes, it is quite normal in that we find similar structured systems everywhere in nature and in products made by humans. But they were only found much later. And this much too rarely. Even more rarely were they understood when they were observed. Therefore, all this is still not common knowledge within science, not in the scientific world, and therefore not in educated circles of the broader population. Above all, when such characteristic structures are found, there is a lack of deeper understanding of how they come about. It is still widely unknown that these are completely different systems, namely, non-equilibrium systems. It was also a long path towards my own realization of this.

While it took a long time for me to make the observations described above, and only gradually did I come to a deeper understanding of them, I had no plan, no expectation of where the research would lead me. I originally had only a very vague idea of the subject of my research. The why-question was what drove me, I wanted to understand the phenomenon fundamentally. Instead of a research plan, I formulated a hypothesis of what might be the cause of an inexplicable observation. Then, I forced myself and my employees to formulate an opposite hypothesis that should not be absurd, but had to sound just as reasonable. I designed at least two suitable experiments for this, one that would confirm the first hypothesis, another that would refute the second. With this, I wanted to prevent myself from falling in love too quickly with the first spontaneously designed

---

[39] More precisely, we should say: "frozen dissipative structures", because Prigogine uses the term to refer to dynamic structures in systems that are actively away from equilibrium, in which something happens; see also https://en.wikipedia.org/wiki/Dissipative_system; the structures I had discovered had arisen in dynamic systems, but, after cooling the mass, they solidified, that is, "froze".

hypothesis and then possibly researching in the completely wrong direction. Whatever came out would thus force me to further alternative hypotheses. I stumbled from one hypothesis and the subsequent surprising (or puzzling) experimental result to the next, never knowing what was waiting for me at the next turn. So, my research shimmied from one accidental observation to the next coincidental one.

I am far from being the only one in the world of research using such a methodology. Unfortunately, we do not know how most researchers came to their key thoughts, how they got the idea for a crucial experiment or the interpretation of an important observation. August Kekulé is said to have seen the structure formula of benzene in a waking dream, as the atoms danced in front of his half-sleepy eyes and arranged themselves into rings: a completely new concept for chemical structures! Wilhelm Conrad Röntgen recognized, after covering the cathode ray tube, that the paper painted with color still glowed: It could be an *invisible* radiation, what a new thought! I had to scribble on dozens of sheets of paper on the floor of a crowded airport building in the midst of hysterical people. I was in a state of complete mental isolation, hardly noticing anything of my surrounding. I didn't care at all whether I got a flight that night—my American colleague and friend who was with me took care of all of that. I was in deep meditation or in a kind of self-hypnosis until the solution appeared before my mind's eye: Not only are the particles not evenly distributed, but they are also quite firmly wetted. In addition, they form tubular, branching pearl chains within a common tubular skin—a previously unthinkable idea!

James Watson had to revise the prejudice that the bases were sitting on the outside in the DNA helix and had to mentally move them inward. He got the initial idea that the DNA structure could be helical from Linus Pauling's discovery of the helix structures in proteins—an unjustified *analogy,* because proteins and DNA are not analogous at all! Nothing fit, until—because he couldn't get any further with his models—he moved the spiral chain to the outside and the bases to the inside. Watson tinkered and painted, and *suddenly,*[40] he realized how the bases could be opposite each other inside the double helix. But he still hadn't realized that adenine didn't sit opposite adenine, and guanine didn't sit opposite guanine, and that such an arrangement wasn't possible. He had to overturn another important prejudice: In the DNA double helix, one chain does not form a helix with another, exactly identical chain, which was the original idea! Here, *like does not pair with like!* So, he made cardboard models of the bases—and found the now known base pairing adenine/thymine and guanine/cytosine.

It must be clear that the *flash of genius,* the *spark of inspiration* does not come from a vacuum in a brain where nothing else is happening, but that these researchers' brains has been occupied, intensively and for a long time, with various hypotheses, observations, calculations, experiments, with the specialized literature of other researchers and in discussions with other scientists, competitors and sympathetic colleagues, and much more. Only then can such flashes occur—now, let us think of *serendipity!* Critics may

---

[40] *J. D. Watson, loc. cit. p. 229.*

here interject: Most such sudden insights as well as most or at least many inventions, are "simply analogies". If it were that easy, many people with similar or even better qualifications could make such "simple" analogies. But, applying or transferring a principle somehow *analogously* to a completely different question, is a creative idea, a flash of genius: You have to come up with that first! And many such flashes of genius are only possible on the basis of a broad and deep knowledge of data and facts.

Improvisation is Coincidence in the Brain

Back to the *flash of genius,* to the *sudden insight.* As a very amateurish piano player, I regret that I cannot play freely, let alone improvise. When we listen to an improvising pianist or saxophonist, we are witnesses to sudden musical inspirations that these artists then translate directly into finger movements or breathing rhythms. An article by Sebastian Kempkens in the weekly newspaper DIE ZEIT with the title *Flourishing Fantasy (Blühende Fantasie)*[41] made me think—beware, analogy!—about the similarity or dissimilarity of improvisation and flashes of genius.

The article began with the already mentioned near-catastrophe "Apollo 13", which was only averted because a number of people in Houston were able to improvise a method by which the astronauts could build a CO_2 filter with objects available on board. In a completely different field, the managing directors of the German company "Spirituosen Manufaktur", which was on the verge of bankruptcy due to the Corona pandemic, improvised the production of fragrant disinfectants because they could no longer sell their special brandies. The pandemic has inspired or forced many people to improvise, to engage in new activities.

In music, especially in jazz, it is known that good, interesting improvisation is only possible if you master the instrument and the music. Those of us who, like me, cannot perfectly operate the piano will continue only to dream of improvisation. I can just play the compositions of my favorite composers that I have practiced for a long time. Neuroscientists have now examined what happens in the brains of musicians when they improvise. They put jazz musicians and their keyboard into a magnetic resonance imaging machine. Pieces that had been rehearsed showed activity in a different area of the brain than the improvisations; in the latter case, the brain areas responsible for self-control were almost deactivated, the brain was in a state similar to that during the sleep phase of intense dreams.[42]

---

[41] https://www.zeit.de/2021/01/improvisation-corona-krisenpolitik-2020-apollo-13-mission-ronald-reagan

[42] Original publication of this research result with a somewhat cumbersome title: K. Dhakal, M. Norgaard, B. Adhikari, K. Yun, M. Dhamala, "Higher Node Activity with less Functional Connectivity during Musical Improvisation", Brain Connectivity, Vol. 6 No. 10 (Oct 2016), 772–785.

This little story fits in with these observations: Paul McCartney has reported that he heard the melody of the song *Yesterday* one morning in 1964 in a dream and wrote it down immediately after waking up.[43] This is very reminiscent of Kekule's dream of the benzene ring!

Let's take a step further and ask: What does a composer do, how did my favorite composer Beethoven compose? I now simply assert—let's call it a little presumptuous—*my working hypothesis:* Composing is nothing other than mental improvisation, which the composer then writes down, and then afterwards—because he can hear the written mental improvised sounds, melodies and rhythms in his head—refines and optimizes step by step.

I thought about this further because I knew that I had many (if not most) of my ideas not in the office, not in the laboratory, but outdoors while hiking, kayaking or during my crane research, and then wrote them down. Or on the toilet (and immediately pulled my notebook out of my shirt pocket). Kekule, in a state between sleep and waking, dreamed, and then painted the benzene ring. I sat on the floor in the chaotic Newark airport and was the only one who was not hysterically looking for a flight, but rather, as in a trance, drew one picture after another until finally one of the hallucinations could explain all of my experimental results.

The young chemist and later Nobel laureate Benjamin List had an idea after his doctorate, one that, as the Nobel Committee formulated it after the announcement, "was so brilliant and simple that many wondered afterwards why no one had thought of it before".[44] His idea consisted of a question: Do you really need a whole enzyme to catalyze a reaction, or does a small(er) part suffice, e.g., only a single amino acid? He tried it with proline. When he started the experiment, he thought: "Maybe this is a stupid idea." It was anything but stupid, it was the beginning of a new chemical science (*Organocatalysis*) and is now already used industrially.

What are such events in the brain that enable the composer to suddenly invent the melody and sound, the fantastic tone sequences that make us listeners breathless, and the researcher to suddenly have the idea, the saving idea that prevents bankruptcy, the craftsman who is missing a spare part for the repair and instead uses a completely different part, which he only has to grind and adapt a little? Improvisation sometimes helps us out of a tight spot, but often, it is life- or existence-saving in crisis mode. Crises are nothing other than serious negative coincidences, accidental events that dramatically disrupt what was originally planned. The improvisation coincidence goes up against the crisis coincidence.

In my opinion, a sudden idea is nothing other than a coincidence in the brain. Far advanced creativity is based on the ability or the gift of individual people to allow such

---

[43] Stefan Klein, "Nobody thinks for themselves", ZEIT-MAGAZIN 4. 3. 2021, p. 16, https://www.zeit.de/zeit-magazin/2021/10/kreativitaet-methode-austausch-geist-wie-wir-die-welt-veraendern

[44] https://www.zeit.de/wissen/2021-10/nobelpreis-chemie-benjamin-list-david-macmillan-asymmetrische-organokatalyse-forschung/komplettansicht

coincidences in the brain more often than the average member of the population does. Of course, it is not *one* coincidence, but rather many, and they become possible because—as neuroscientists have determined—the brain of the improviser has switched out of the self-control mode and into an almost dream-like mode. Now, the neurons are firing in an uncontrolled way, but the brain is prepared.

Even though what I have presented here is speculative, it is at least in line with the research findings on how the brain works. Hermann Haken and his daughter Maria Haken-Krell published a book in 1997 entitled "Brain and Behaviour" (*Gehirn und Verhalten).*[45] Haken is the founder of the theory of *Synergetics,* to which I will return later. In the aforementioned book (and in original publications cited below), the two authors examine the processes in the brain from the perspective of this theory. Based on investigations, for example, of the conscious control of our movements, they compare the results with the control of motors by computer programs: There is no room here for fluctuations and only then delayed reactions if they are programmed, and then in a deterministic way. However, in the experiments described by the authors, surprising—and unconscious!—changes in the course of movement of the fingers or legs, which suddenly moved synchronously, leading them to the conclusion: "In a certain way, the brain has one—or maybe many—chance generators." Consequently, the title of their Chap. 7, *Chaos im Gehirn (Chaos in the Brain),* and that of Chap. 9, *Gehirnströme spielen verrückt (brain waves go crazy).* Here it is explained that the brain works close to points of instability and the currents form superimposed spatial patterns.

Experiments carried out as early as 1965 are particularly interesting. In these experiments, each person was asked to raise their index finger whenever they wished. The book then continues: "After a while, she actually moves her finger. But, and this is the crucial discovery, about sixty thousandth of a second *before* this action, special electrical potentials in the brain can be observed with the help of the EEG. It is as if the brain were preparing the action." According to the authors Haken/Haken-Krell, this is an act of self-organization[46] in the brain, in this case, probably as a combination of material (chemical) and electrical processes.

This reminds me of how I play soccer as a goalkeeper. I've been a passionate soccer player since childhood and have particularly distinguished myself in the goal. Not that

---

[45] H. Haken, M. Haken-Krell, "Gehirn und Verhalten—Unser Kopf arbeitet anders, als wir denken", DVA 1997 (German), pp. 10 ff. and 125 ff. as well as p. 258 f. For English sources, cf. http://www.scholarpedia.org/article/Self-organization_of_brain_function. My considerations are also in line with the research results based on EEG and fMRI studies by J. Kounios and M. Beeman, which they describe in their book 'Das Aha-Erlebnis. Wie plötzliche Einsichten entstehen.' dva 2015 (German). For original papers, see Kounios and Beeman, "The cognitive neuroscience of insight", Annu. Rev. Psychol. 2014; 65: 71–93; https://pubmed.ncbi.nlm.nih.gov/24405359/

[46] We will come back to the phenomenon of self-organization several times later in the book, including under the term "order from chaos", and deal, in particular, with the findings of Ilya Prigogine and Hermann Haken and their closer colleagues.

I've ever played in proper clubs or even in a higher league, no, it was always recreational soccer, but I was and am playing very often, very intensely, and also often against much stronger teams, so that I was challenged even more than usual. Soccer is one of my favorite sports for freeing my brain and becoming creative again. For decades, we had a regionally successful company soccer team, and I played soccer in China for thirteen years two to three times a week with Chinese people who were much younger than me. Back in Germany, I'm now playing again in an old-timers' team. Before the game starts, I let myself be warmed-up as a goalkeeper: I ask my teammates to shoot one ball after another at my goal, easy shots at first, then increasingly difficult ones and many in quick succession from all directions. In this way, I want to turn off my conscious brain (Kahneman's *System 2*), I want to achieve that my brain controls my reactions without my conscious control. I want to force my brain to stop thinking about the current company problems or the burning research question—or this book here—by focusing on the many balls. This allows my brain to tell my arms and legs what to do without my conscious control. If I have succeeded in this, if I am—as one now also says in modern German—*in flow*, then I am particularly good, particularly reaction-fast, and often I cannot remember myself how I fished out the one or the other ball that my teammates "already saw in the goal". These reflex reactions are also, same as the improvisations of jazz musicians, *coincidence in the brain.*

As this chapter concludes, I would like to avoid a possible misunderstanding: Great ideas, flashes of genius and inspirations are not enough. They are, by far, not everything in research, in art, in engineering, in entrepreneurial innovation. Before and after any sudden insights, there is intensive, systematic, clean and often very time-consuming work, which can be quite frustrating in between moments of triumph, because the real chemical and physical experiments do not run as the thought experiments had promised. The realization of great ideas always requires intensive thinking, but this must be followed by regular breaks in relaxed situations, otherwise, no research results, no work of art, no new product will come from the idea. Even the best ideas require a lot of patience afterwards. And, in addition, a great deal of tenacity, a highly developed resilience against rejection, humiliation, sometimes massive and aggressive attacks; and against the constantly nagging self-doubts. How much tenacity the founders of BioNTech, the married couple Uğur Şahin and Özlem Türeci, the first pioneers of mRNA research, how much patience, long-term thinking and far-sighted scientific work together with entrepreneurial thinking and the patient continuation of work even at low points, had to muster, we can only guess from the outside.

The Chinese have kept the following beautiful Cheng Yu from ancient times to describe someone who pursues a goal with unyielding determination:[47] 精卫填海。(jīng wèi tián hǎi) "Jing Wei fills the sea."

---

[47] Text taken from the proverb collection of the East Asia Institute: https://bit.ly/3IKEQQy

Nǚwā (女娃), the daughter of (mythical) Emperor Yàndì (炎帝), played on the beach of the East China Sea and saw the deep blue sea, which tempted her to swim in it. But when she had moved away from the shore a little, the sea became stormy and huge waves piled up. The sea god had allowed himself a joke with Nǚwā, who could not manage to get to the shore and drowned. When Yàndì heard about it, he hurried to the sea and called the name of his youngest daughter, but all he found was a small bird that fluttered around him and called out shrilly. Yàndì was sure that Nǚwā had transformed into this bird and cried bitter tears. Then, the bird suddenly called "jīngwèi", flew to the West Mountains, picked up a stone there and threw it into the East China Sea. From then on, the little bird flew back to the West Mountains again and again, picked up stones and threw them into the East China Sea. The sea god understood what the little bird was up to, laughed at it and said: "The sea is so wide and deep that you will never be able to fill it." But the bird replied: "I will never stop, and one day—believe me—I will have filled the sea."

(Photo credit NASA https://www.nasa.gov/images/content/190026main_filterbox.jpg).

Here is the original photo of the CO_2-filter in Apollo 13 that had to be improvised in a very short time, otherwise, the astronauts would have died. They could only use parts available on board. Here—https://spacecentre.co.uk/blog-post/story-apollo-13/- you can find a guide on how to build such a filter yourself based on this example.

More details on contents in this chapter in Appendix 2, Part 2, Appendix 5, 6, 7, 11 and 12, Part 1.

"Equilibrium is good, Non-Equilibrium is bad"—is that True?

4

Abstract

The widely held views of *equilibrium* are discussed, including the concept of *dynamic equilibrium (steady state)*. Entropy is explained in an easily understandable way, as is, despite the 2nd law of thermodynamics ("continuous increase of entropy in the universe"), complex structures can arise. Thus, we learn the basics of non-equilibrium thermodynamics. We understand that we ourselves and everything around us are non-equilibrium systems with dissipative structures. Otherwise, mayonnaise would not be stiff. "For organisms, *equilibrium* means death and decay." (Ludwig von Bertalanffy, founder of the *steady state (or: dynamic equilibrium)* concept).

Do not let the term *thermodynamics* make you uncomfortable in this moment. I will not try to teach you thermodynamics, forget the word, just cross it out in the book if it occurs. Let's just talk about *equilibrium* and *non-equilibrium* and think about both.

We all would like everything to stay nicely in equilibrium. The economy, ecology, climate, even the mutual atomic threat, in German, interestingly called the *balance of terror ("Gleichgewicht des Schreckens")* (no doubt for the purpose of calming our fears). While in the US, it's known as *mutually assured destruction*, an idea also based on the concept of equilibrium, the so-called *Nash equilibrium*...[1] We read the admonishing articles by journalists who express growing concern when, once again, currencies, the stock market, the financial economy, the real estate market have "gone out of balance (or "equilibrium")". And all the more so in regard to the climate. Or we hear about problems with the *ecological equilibrium,* because nature's biggest predators are going missing. In an article

---

[1] cf. https://en.wikipedia.org/wiki/Mutual_assured_destruction and https://en.wikipedia.org/wiki/Nash_equilibrium

© The Author(s), under exclusive license to Springer Fachmedien Wiesbaden GmbH, part of Springer Nature 2023
B. Wessling, *What a Coincidence!*, https://doi.org/10.1007/978-3-658-40671-4_4

on spektrum.de, one reads: "*The earth is what makes our life possible at all—still, ... climate change, resource consumption, population development, energy use or pandemics threaten [the planet's] sensitive equilibrium to our disadvantage.*" [2] Apart from the fact that the destruction of biodiversity is, as usual (and in spite of its extremely critical significance), missing from the list, the quote illustrates our desire to maintain *equilibrium*. It has even manifested itself in the Constitution of the Federal Republic of Germany which speaks about the "requirements of macroeconomic equilibrium" in Article 109.[3] In practice, however, it sounds like this: If the economy stagnates or even shows *negative growth* (a great term!), it is out of equilibrium. But if it grows, the requirements of equilibrium are met—what a strange idea of equilibrium!

It is time to clear up some confusion about this: None of these things have ever been in equilibrium. What we so readily take for *equilibrium* is, at best—as some experts like to say—a *dynamic equilibrium* (also called "steady state"), or at least something that, during our relatively short memory span, looked and felt like equilibrium. The creator of the steady state (or: dynamic equilibrium) concept was Ludwig von Bertalanffy.[4] When I read the words *dynamic equilibrium* (and especially the original German term "Fliessgleichgewicht", to be translated as: "flow equilibrium"), I imagine: Something— substances, money, energy—flows into somewhere at a certain rate per unit of time, and something—again substances, money, energy?—flows out again at about the same weight or volume. A river *flow* has the same water level while we are watching it, but that does not mean it is in equilibrium, but rather in a state often referred to as *dynamic equilibrium*. In my opinion, this is not only misleading, but even deceptive; more on that later.

First of all, let's take a look at what is meant by *equilibrium* in thermodynamics. We take two very familiar substances, sugar and water, and pour them into a glass. If we look closely and measure carefully, it becomes clear: When sugar dissolves in water, heat is released. The *heat of solution (or: enthalpy of solvation)*. The heat released is an indication that we are observing a process on the way to equilibrium. In a second experiment, we use concentrated sulfuric acid and water, and notice: *a lot* of heat is released! But be careful if you want to repeat this experiment on your own: You must put the water into your glass vessel first, then you add the acid. We chemists have a saying, which, for German chemists, is a rhyme: "Erst das Wasser, dann die Säure, sonst geschieht das Ungeheure." The meaning of it is: If you wrongly add the water to the acid, "something monstrous will happen". As it turns out, Anglo-American chemistry students also memo-

---

[2] https://bit.ly/3gnsG3B (spektrum.de).

[3] Article 109, paragraph 2 of the German Constitution reads (translated): "The Federation and the Länder [States] jointly fulfil the obligations of the Federal Republic of Germany under acts of the European Community for the purpose of complying with the budgetary discipline and, in this context, take into account the **requirements of the macroeconomic equilibrium**."

[4] cf. https://en.wikipedia.org/wiki/Ludwig_von_Bertalanffy, https://en.wikipedia.org/wiki/Steady_state and https://en.wikipedia.org/wiki/Open_system_(systems_theory)

rize this warning with a rhyme: "First the water, then the acid, otherwise it won't be placid." In some places, they teach it in a different way: "Do like you oughta, add acid to water." Because, if you pour water onto concentrated sulfuric acid instead, heat is suddenly released on the surface of the acid, where the water touches the concentrated acid, so that part of the water vaporises explosively, taking sulfuric acid droplets with it and endangering your face, your eyes, or at least your clothes.

In general, there are no exceptions: All substances, all systems strive for equilibrium. This is because, on the one hand, the energy content of every system tends towards a minimum, and, on the other hand, the entropy tends towards a maximum. I will explain the term *entropy* in the next subchapter. Even systems or processes for which no heat is released cannot prevent themselves from moving towards equilibrium if only enough entropy is generated. That is sufficient, and the relevant system will inevitably move into the equilibrium state. Unless something else happens. More on that later.

But what is *dynamic equilibrium*? The Austrian-Canadian chemist Ludwig von Bertalanffy introduced this term in a monograph from 1953 and worked it out in depth.[5] He focuses on organisms as open systems. This refers to systems that exchange matter and energy with their environment. He explicitly relies on Ilya Prigogine's non-equilibrium thermodynamics, which we will come to later. Such systems, such as the continuously operating reactors considered by Bertalanffy, in particular, living organisms as well as complex communities and ecosystems, can "under certain conditions reach a time-independent state in which all macroscopic variables remain unchanged, although macroscopic processes of import and export continue. Such a state is called *Fliessgleichgewicht* [i. e., 'steady state' or 'dynamic equilibrium']."[6] The condition that all macroscopic variables do not change may apply quite well to a chemical continuous flow reactor to which the same amount of starting material is fed day in, day out, whose temperature is kept constant through controlled heating and cooling, and from which the same amounts of product and waste are removed day in, day out. And that can be done for a whole year until the plant is shut down for routine inspection and maintenance and restarted after a few days or weeks.

This is hardly the case for an organism in a comparable way: In which organism do "all macroscopic variables remain unchanged" when we consider only a single day's course? Neither the supply of food is continuous, nor the removal of products, in this case, the excretions. Nor does the temperature remain the same during the day, let alone

---

[5] Ludwig von Bertalanffy, "Biophysik des Fliessgleichgewichts," (German) Vieweg (Braunschweig) 1953. In the following, I translate and quote from the 2nd (revised, updated and extended) edition worked out by Walter Beier and Reinhard Laue initially in cooperation with Ludwig von Bertalanffy, who, however, died before the work was completed, Akademie-Verlag Berlin 1977. An English edition has been published here: Scientia 48 (89):361 (1954), "The biophysics of the steady state of the organism"; in this book, he also outlined his "dynamic equilibrium"/"steady state" concept: "General system Theory", see https://www.academia.edu/38207367/Von_Bertalanffy_Ludwig_General_System_Theory

[6] L. v. Bertalanffy, W. Beier, R. Laue, translated from loc. cit., p. 54/55.

do organisms remain unchanged in their macroscopic sizes over the course of a year: They grow, they age; the fur, the hair, the antlers grow and are later shed or wear out—and many other macroscopic changes. Nevertheless, Bertalanffy and co-authors also refer to living organisms as typical representatives of *dynamic equilibrium (steady state).* At the same time, they are fully aware of the fact that this state is just *not* equilibrium. On page 55 (updated German edition), they continue to write that, in contrast to equilibrium processes, which are reversible, the processes in dynamic equilibrium are irreversible. In order for an open system like an organism to be operational—or, in my words: alive—it needs "to maintain its deviation from equilibrium" through constantly supplied energy.

In another publication, Bertalanffy becomes even clearer. He writes: *"Biologically, life is not maintenance or restoration of equilibrium but is essentially maintenance of disequilibria, as the doctrine of the organism as open system reveals. Reaching equilibrium means death and consequent decay."*[7] (Thermodynamics does not use the term "disequilibrium", but rather "non-equilibrium", the sense is essentially the same.) In other words: The term *dynamic equilibrium (steady state)* does not simply mean another form of *equilibrium,* but, in reality, a *non-equilibrium system,* that is, simply the opposite. Although these are *relatively stable* systems for a certain time, they are nevertheless non-equilibrium systems. To clarify it all and repeat what Bertalanffy wrote here: *Equilibrium* means death and decay for organisms. Nobody can seriously disagree with this, because it is a fact, a natural law.

Back to the *macroeconomic equilibrium* in the German Constitution: In any case, this term is not in line with the definition of *dynamic equilibrium,* in which macroscopic variables should not change at all. If they change, the system is no longer a "steady state" or dynamic *equilibrium* system, which is also a non-equilibrium system, but simply and plainly a typical non-equilibrium system. A system that grows. Qualitatively spoken, not much different from one that stagnates or shrinks, everything is always *non-equilibrium.*

That is why I find the term *dynamic equilibrium* ill-chosen, and the other English term is even more misleading: *steady state,* that is, literally interpreted, some system which constantly stays the same. Both terms suggest something that everyone would like to have, but which—objectively speaking: fortunately!—does not exist as long as life goes on and the universe continues to develop and change. So, what distinguishes equilibrium from non-equilibrium?

---

[7] Ludwig von Bertalanffy, General Systems Theory, Verlag George Braziller, N. Y. 1968, p. 191; "Biologically, life is not maintenance or restoration of equilibrium, but is essentially maintenance of disequilibria, as the doctrine of organism as open system reveals. Reaching equilibrium means death and consequent decay." (see https://www.academia.edu/38207367/Von_Bertalanffy_Ludwig_General_System_Theory)

"Entropy": Heard of it, but didn't Understand it?

I wrote above that all systems tend towards an equilibrium state, unless "something else" happens. And now, we will look at what "something else" is. It's about entropy. Please, no worries, actually, it's quite simple. At least, as long as you don't want to become a thermodynamicist, and I don't want to convince you to be that. You can imagine entropy in two ways. On the one hand, as a description of disorder or order, which is called the *statistical interpretation of entropy:* A system that, after a process that we observe, shows less order than before has experienced an increase of entropy.

On the other hand, we look at the dimension (the unit) of entropy. This is "heat unit per temperature unit" ([J/K], "J" = Joule, a unit of energy, "K" = Kelvin, the unit of temperature). So, entropy has something resembling an energy, but it is not really a true form of energy. Entropy is responsible, for example, for the fact that power plants can not work with 100% efficiency. I therefore refer to it here, for our purposes, as *waste energy* or *energy loss.* This is, although entropy is not referred to as such anywhere else, therefore not inappropriate, because one thing is clear: You can not collect or recover entropy in a system, not to mention use it for anything. Entropy is simply useless, or, to be more precise: The higher the entropy content in an energy form, the less we can use this energy. This may bother us, it may be annoying, but it is so, it is a law of nature. In addition, there is not a single process associated with matter in the world, in the entire universe, that is not associated with entropy production (or: generation). No matter what happens, no matter what you do, no matter what the weather is doing or the nuclear fusion in the sun: Entropy is created.

For centuries, it was the ambition of numerous inventors and tinkerers to construct a perpetual motion machine, that is, an apparatus or machine that does something endlessly without any input of energy. All such attempts have failed. There is no windmill that can generate the wind with which it operates and, in doing so, generate the wind that drives it again. Because, for example, entropy is always generated when operating a machine or process, that is, more primary energy is required than is made available in form of usable energy or is made available to us by the machine.[8]

Our brief explanation of entropy above has two consequences: If we use the statistical interpretation of entropy, then an equilibrium state is identical to the maximum possible distribution of substances (or: system elements) in each case. And further: In a system, we can only *reduce* the entropy if we or someone or something else act with a very

---

[8] For the same reason, I am very sceptical of the repeated efforts that are being made to recover CO_2 from the air in order to stop or at least slow down climate change: With recovery processes, more entropy will be generated than can be reduced by the recovery of entropy that is distributed everywhere in the form of CO_2. There are only two ways to reduce CO_2: On the one hand, by growing plants and regenerating bogs, and on the other hand, by wasting and consuming less primary energy.

high energy input. With a *really extremely* high energy input. When energy is converted, entropy is generated again, but outside the system in which we want to reduce entropy by importing energy. Attention: I wrote very deliberately that "energy is *converted*", not *generated*! We cannot *generate* energy, at least, *we* can not. The universe could do it at the very beginning, now the energy is stored in all kinds of substances, whether it is coal or oil and gas, or uranium, which provides energy when it is split, or even hydrogen nuclei, which fuse themselves to helium in the sun.

This has a third consequence: If we would like to reduce the disorder and create more order in a system, that is, reduce the entropy and build structures, we therefore have to expend energy and introduce it into said system. This results in the fact that we have increased the entropy where we get the energy from, that is, we export entropy from the system that interests us to somewhere outside this system. However, in addition to the energy required for this, entropy has also been generated, so that both contributions together, the increase in entropy where energy is converted and the export of entropy from the system we are considering, result in an in a total higher entropy value than before.

This is how our industrial production works: We start with relatively unstructured raw materials, for example, iron ore, apply energy-intensive chemical and then physical processes (both of which use coal), and the result is a precisely shaped steel pipe.[9] In the steel pipe, the entropy is now lower than in the iron ore. But in the place from which we got the iron ore, we created a lot of disorder: The landscape is destroyed, entropy has increased enormously there. The same is true in the place from which we got the coal. We can see it in the spoil heaps. And after we converted the coal into industrially useful energy so that we could produce a steel pipe from iron ore, we also left behind a lot of ash from the combustion of the coal. Not to mention the CO_2 that is now distributed throughout the atmosphere. The destroyed landscape in the iron ore mine, the coal spoil, the ash, the factory effluent, the CO_2 distributed in the atmosphere are all the visible or measurable indications of the entropy increase with which we have to pay for the entropy reduction in the steel pipe.

Entropy does Always Increase? Yes, but not Everywhere!

You have certainly already heard and read: In the universe, entropy does inevitably and irreversibly constantly increase. That is correct. However, all too often, it is not differentiated, with the misunderstanding being spread that entropy—because it always increases

---

[9]For those interested in more detail: In the ore, iron is not present in metallic form, which we would then simply have to melt and form to obtain the pipe, no. In the ore, iron is present as iron oxide; it must be chemically reduced, which is also only possible with high energy expenditure, and only then do we get metallic iron, which can be melted and shaped it into a pipe.

"in the universe"—is said to *always* tend *everywhere* towards higher values. That is not correct, as I have shown with the example of the steel pipe: The entropy always increases overall, but this is the case in the mining area, in the spoil heap, on the landfill, where the ash of the coal-fired power plant is stored, in the warmed-up and discharged cooling water of the factories, in the atmosphere of the earth and in the universe (the destination of the entropy that we have emitted as infrared radiation), but not in the steel pipe.

In the cases we considered in the previous chapter, when, to put it simply, we made a stiff mayonnaise from egg yolk, lemon juice and olive oil, all three of which are liquid, the entropy in the mayonnaise decreased. Complex micro- and nanoscopic structures were built there that make the liquid emulsion viscous; we looked at these structures in SEM photos. Outside the mayonnaise, entropy has increased: In our kitchen waste bin, there are now quite useless eggshells, a squeezed lemon, and, if we do this more often, an empty glass olive oil bottle, all indicators of an increase in entropy. We also either stirred the emulsion vigorously ourselves, thereby generating and radiating body heat and exhaling more CO_2 than usual. Or we used an electrically operated kitchen appliance, for which the corresponding entropy increase took place somewhere far from us in the power plant and its environment. Another entropy contribution arose from the operation of the appliance itself, which does not work with 100% efficiency, and therefore also generates entropy, which we can partly feel on the basis of the heat development in the appliance.

Let's go back a little and look at the complex structures of a chicken egg and a lemon: How did they come about? Well, we can't look at that in detail here, but ultimately, it was solar energy that indirectly generated the food for the chickens, and the same energy source was responsible for the growth of the lemon tree, its leaves and flowers, for the flight of the busy bees that pollinated the flowers until, finally, the lemon was ready. Without the sun, nothing grows, and the bees don't fly, as we notice in winter.

Ilya Prigogine[10] is the father of non-equilibrium thermodynamics and was awarded the Nobel Prize for it. He found out (describing it qualitatively and quantitatively) that, in open systems, when an *overcritical* amount of energy enters the system, and the system is thus *far from equilibrium*, structures can spontaneously arise by self-organization. So, entropy decreases. Where there was chaos before, with everything whirling around, suddenly, highly complex order arises. He called this *dissipative structures:* They arise through self-organization. You will understand the term *dissipative* in a moment.

The *Rayleigh-Bénard convection*[11] is such an open system, open because it can at least exchange energy with the environment. Most other open systems also exchange matter with the environment. The *Rayleigh-Bénard convection* is not a chemical reaction, but a rather simple physical process: In a flat glass dish, silicone oil is heated. This does not

---

[10] https://en.wikipedia.org/wiki/Ilya_Prigogine

[11] http://www.scholarpedia.org/article/Rayleigh-B%C3%A9nard_convection

sound very exciting yet. We can add fine copper or other metal pigment powders to the oil, although just for better visibility, they are unimportant for the principle. Now, we slowly increase the heat input, and we will notice that, at a certain temperature difference, pentagonal and hexagonal honeycomb cells suddenly form, as the photo shows.[12]

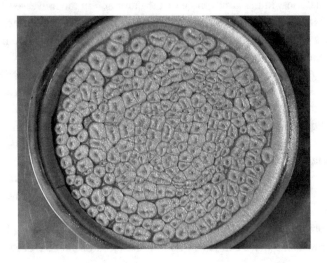

What happened there? With a supercritical heat input from the heating plate into the glass dish, the oil can no longer cope with the difference between the temperature at the bottom in contact with the heating plate and the one at the free surface (can not "dissipate" the enormous amount of energy) in any other way than by synchronous rolling, whereby the liquid rises in the center of the cells and falls on the boundaries between adjacent cells: *dissipative structures.* In this way, the excess heat can be dissipated much more efficiently to the outside!

According to what we have discussed earlier, the explanation is clear: While the oil-/ pigment suspension was structureless before heating and also at the beginning of our experiment—that is, it contained a lot of entropy, all particles and oil molecules were statistically evenly distributed—suddenly, some complex order was forming: The fluid molecules move in a synchronous rolling motion, along with the pigment powder particles. There is no other interpretation: The entropy has suddenly drastically dropped! How can that be? For our purpose in this book, we understood entropy as a kind of *waste energy* or *energy loss*—the supercritical energy import by the heating plate causes an entropy export, an *energy loss export* by massively increased convection made possible by the dynamic honeycomb structures. Because, otherwise, the system would not be able

---

[12] https://www.experimente.physik.uni-freiburg.de/Thermodynamik/waermeleitungundkonvektion/ konvektion/benardkonvektionszellen; Reproduced with kind permission of the Faculty of Physics of the University of Freiburg.

to cope with the supercritical energy input. Dissipative structures form, as they were first called by Ilya Prigogine. He had also recognized that self-organization is associated with supercritical energy import and entropy export.

If you look at photos of the solar surface, you will recognize the same phenomenon there! Only there, the *Bénard cells* are a little bigger than in the flat glass dish in our laboratory, namely, more like the size of the earth or even larger.

Prigogine dealt very intensively with *oscillating chemical reactions.* Normally, reactions take place in one direction: They start with at least two raw materials, from these, an end product is generated, and perhaps there are by-products. From a thermodynamic point of view, the reaction has come to equilibrium, it is finished. Oscillating chemical reactions are rare and complicated, but still simpler than what happens in the cells of all organisms. And that was the reason why Prigogine dealt with it. The simpler chemical oscillations could serve, for example, as models for the citric acid cycle[13] in our cells.

Let's look at the perhaps most famous oscillating reaction in a homogeneous phase, the *Belousov-Zhabotinski-Reaction.*[14] It oscillates—a fact that is visible due to suitable indicators—between different states, and this leads to impressive dynamic patterns in an execution in a flat dish.[15] Here is a photo of a typical pattern (taken from a video).[16]

---

[13] https://www.britannica.com/science/tricarboxylic-acid-cycle

[14] You will find a video of the reaction sequence in a glass cup under the instructions for carrying out the experiment on this page: http://www.chem-page.de/?view=article&id=69:belousov-zhabotinsky-reaktion.

[15] https://www.youtube.com/watch?v=i3g5jB0ddQM

[16] https://www.flickr.com/photos/nonlin/3572095252/in/album-72157623568997798/ with photos and videos showing the development of the reaction; printed with kind permission of Stephen Morris (Univ. of Toronto, Canada) and Mike Rogers.

The reaction can run for a few minutes without replenishing consumed materials. The reaction mechanism is relatively complicated. In three different reaction groups, there are a total of 18 partial reactions. You can find good introductions and overviews here.[17] Theoretical descriptions of these reactions require coupled differential equations. The energy supplied comes from the *chemical potential* contained in the reactants. The entropy export takes place through the formation of CO_2, which leaves the reaction space: So again, local structure formation is linked to entropy reduction in the relevant system, and entropy increases when looking at the entire universe.

Further Research on the Phenomenon of "Self-Organization"

Prigogine was probably the first, but not the only, researcher to turn to the phenomenon of *order from chaos* or *self-organization*. He was, above all, the one who combined it with a completely new theory of non-equilibrium thermodynamics. Other researchers have approached this topic from other perspectives.[18] One of the outstanding ones among them was Hermann Haken, whom we already met in the previous chapter. He is the creator of the theory of *synergetics*. He defines it as follows:[19] *"Synergetics is an interdisciplinary field of research that deals theoretically and experimentally with systems made up of many (equal or unequal) parts. These are open systems into which energy, matter and/or information flow in and out* [I will comment on this specifically at the end of the quote]. *The focus is on such processes in which the macroscopic state of the system changes qualitatively through self-organization (emergence of new qualities). Synergetics is looking for principles that are independent of the specific type of parts. These parts can be atoms, molecules, photons, biological cells, neurons, people in society, companies in the economy, etc."* [After *"and/or information"*, in my view, it should be added *"and from which entropy is exported"*. The fact that this is missing has probably to do with that Haken and Prigogine knew each other and met at conferences, but obviously

---

[17] http://www.scholarpedia.org/article/Belousov-Zhabotinsky_reaction   or   http://www.chem-page.de/?view=article&id=69:belousov-zhabotinsky-reaktion

[18] For an overview of the wide range of activities, see, for example, R. Feistel, W. Ebeling, "Physics of Self-Organization and Evolution", Wiley-VCH 2011, Preface, as well as pages 27 and 33, see https://onlinelibrary.wiley.com/doi/book/10.1002/9783527636792

[19] translated and cited from H. Haken "Entwicklungslinien der Synergetik" in "Forschungsberichte der Abteilung für Psychotherapie der Universität Bern" 2014, No. 14-I, (German), available online here: https://www.academia.edu/7776383/Hermann_Haken_Entwicklungslinien_der_Synergetik_in_German_edited_by_W_Tschacher_3_3MB, Quote from p. 18. For English sources, cf. H. Haken "Lines of development of Synergetics", https://link.springer.com/chapter/10.1007/978-3-642-67592-8_1, H. Haken "Synergetics", https://link.springer.com/book/10.1007/978-3-662-10184-1, H. Haken "The Science of Synergetics", Van Nostrand Reinhold, New York 1984

were engaged in active competition. Haken was of the opinion that entropy is not the crucial quantity needed to understand structure formation in non-equilibrium systems.[20]] I would like to emphasize that Haken formulates here a very important insight: Through self-organization, we can *observe the emergence of new qualities*—that is, the appearance of new laws when new structures are formed! We should remember this in later chapters.

In fact, the two theories complement each other. What Prigogine calls *dissipative structures* is called *self-organization* by Haken. Among other things, Haken's *information* flows into or out of open systems, a fact that we can safely equate with Prigogine's representation of entropy production and its export. Nevertheless, the respective descriptions are different, with Prigogine working strictly thermodynamically and Haken mathematically. However, the two theories are, even together, not sufficient. They explain *in general* the characteristics of open non-equilibrium systems and the spontaneous emergence of predominantly dynamic structures, where before there was chaos or simply structurelessness. But, for every system that one really wants to understand deeply, one must find out what kind of interactions generate the structures and the dynamics. The general equations show that self-organization occurs in a variety of systems according to the same *abstract* principles. This shows us that it is a general principle of nature: In open systems, a supercritical amount of energy is imported, system components interact with each other in a specific way, and, above a critical point of instability of the system, structures emerge and massive amounts of *entropy* are exported, that is, *information* is generated within the system.

Self-Organization, Structure Formation in Non-Equilibrium—Structure Loss in Equilibrium

The principle of structure formation through self-organization, the export of entropy in open systems far from thermodynamic equilibrium, by importing supercritical amounts of energy, may now be understandable. Entropy reduction is identical to structure formation, with the emergence of information. Please think back to the statistical interpretation of entropy: High entropy is synonymous with disorder, low entropy with order and complex, dynamic structures, that is, a high information content.

When systems strive for equilibrium, they experience an increase in entropy, which corresponds to a breakdown of structures, a loss of information. For example, crystals are dissolved, the individual molecules that were previously ordered in the crystal are distributed evenly, there are no structures. The dissolution of sugar in water is an example of this: Each sugar molecule is surrounded by several water molecules. In a cup, even in a very large one, you will find the same amount of dissolved sugar at all locations after enough time has passed. Gravity does not cause a gradient in solutions, even if we

---

[20] H. Haken in "Entwicklungslinien …", see previous footnote, p. 19.

give the solution a lot of time. Soil erosion, the weathering of rock transported by rivers in the form of sand and silt into the sea, the decay of ancient buildings and sculptures, and also the collapse of societies and cultures on the way to a structureless equilibrium state are associated with an increase in entropy, the loss of information.

Let's look back to the beginning of this chapter. There, I claimed that the financial, currency and economic systems had never been in equilibrium, and could therefore not *get out of equilibrium*. Now, let's take the principle with which we described systems far from equilibrium, according to which structures can form there—does this description apply to economic systems?

We conduct a thought experiment: We stop pumping energy into the economic system. WAIT, some now may call out, have we been pumping energy in so far? Of course we have. We take the solar energy that provides for all sorts of things free of charge, rain, heat, the growth of plants that we and our animal fellow inhabitants of the earth eat, as well as for wind and water power—without the sun, no water cycle on earth, without water cycle no water power, without the sun, no weather that provides wind. The sun radiates an average of about 230 W per square meter onto the earth's surface; the entropy export radiation is just as high, but in the form of unusable *waste energy*.[21]

We humans also take quite a bit of stored energy out of the earth, be it coal, oil, natural gas or uranium. All of this energy storage was filled as a result of energy expenditure in the long history of the universe and the earth—ultimately, always from solar energy; even uranium is a product of death processes in extremely heavy stars/supernovae. This energy expenditure, which was associated with the provision of energy resources now available to us humans, led to a massive increase in entropy in the universe.

Now, once again from the beginning: We are no longer pumping energy into our macroeconomy, only in our thought experiment. "No energy" is only an assumption in absolute terms in order to illustrate the principle. The fact that, while finishing the German edition of this book the delivery of natural gas from Russia to Germany has stopped at all because of the war in Ukraine, and that the risks following from that had suddenly become obvious, clearly shows that economic systems are not equilibrium systems. Because what would happen if really *no* energy were pumped into the economic system at all? We would die. The infrastructure that was built up over years, decades and centuries would decay: cities, villages, buildings, transportation networks, industry, agriculture, everything. Economy and society would gradually approach the real equilibrium state: decay and death. But was entropy exported during the entire time before, while the economic systems were being built up? Yes, of course, on the one hand, into our rubbish

---

[21] W. Ebeling, R. Feistel, "Self-Organization in Nature and Society", Conference "The Human World: Uncertainty as a Challenge", Frolov Lectures 2017, accessible here: https://www.research-gate.net/publication/316878591_Selbstorganisation_in_Natur_und_Gesellschaft_und_Strategien_zur_Gestaltung_der_Zukunft, see also R. Feistel, W. Ebeling, "Physics of Self-Organisation and Evolution", Wiley-VCH 2011, p. 97/98.

dumps, waste mountains and into the sea, into which an enormous amount is poured that does not belong there. On the other hand, particularly through the reduction of biodiversity,[22] because species extinction corresponds to the breakdown of structural diversity, and thus information loss which equals increase in entropy. And thirdly, entropy was and is exported from our economic systems in the form of CO_2, which is blown into the atmosphere, and infrared, which is radiated from the earth into space.[23]

The same applies to our ecosystems: Only as long as the sun sends supercritical amounts of energy to the earth will the structure and dynamics of ecosystems, and thus also our civilization, be built up and maintained. And this is how long evolution will continue, in whatever direction. But we also have to be aware that it is definitely impossible for evolution to practically "go backwards". It could never happen that birds evolved exactly back into the dinosaurs, as we know them from excavations. Evolution is irreversible, just like all non-equilibrium processes.

No Two Snowflakes are Really Identical

All structures on earth, in the solar system and the entire universe are results of *self-organization* in non-equilibrium systems, i. e., dissipative structures. This also applies to complex non-living structures. Let's take a look at snowflakes. There is no one that is like the other! In the picture on the following page, six different basic types are shown.[24] The thesis that no two snowflakes are really identical was put forward by W. A. Bentley in an article in 1922 with the words "no two snowflakes are alike".[25] Of course, such a claim can never be positively proven, because one cannot examine and compare all snowflakes in the world from all times. It can only be assumed to be plausible. This is solely because a snowflake, with its weight of 1 mg, contains approximately 10^{19} water molecules. There are therefore practically infinite possible arrangements of them in a flake. In 1988, the snow researcher Nancy Knight collected and photographed two snow crystals under extremely demanding conditions. During a flight between two cirrus cloud

---

[22] Matthias Glaubrecht, The End of Evolution: Man and the Destruction of Species; C. Bertelsmann 2019.

[23] According to Ebeling/Feistel, a. a. O. S. 87, the earth radiates 1 $W/m^2{*}K$ of entropy into space.

[24] K. Libbrecht, https://arxiv.org/abs/1910.09067, reprinted with kind permission from K. Libbrecht

[25] William A. Bentley, The Guide to Nature (1922); for decades, Bentley, the farmer and naturalist, was thought to be the first person to have taken micrographs of snowflakes (in 1885); it was only much later that the photographs of the German Dr. Johann Heinrich Flögel were found, a lawyer and natural scientist from Schleswig-Holstein, which were taken as early as 1879, see https://en.wikipedia.org/wiki/Wilson_Bentley and https://de.wikipedia.org/wiki/Johann_Heinrich_Ludwig_Fl%C3%B6gel (German text only).

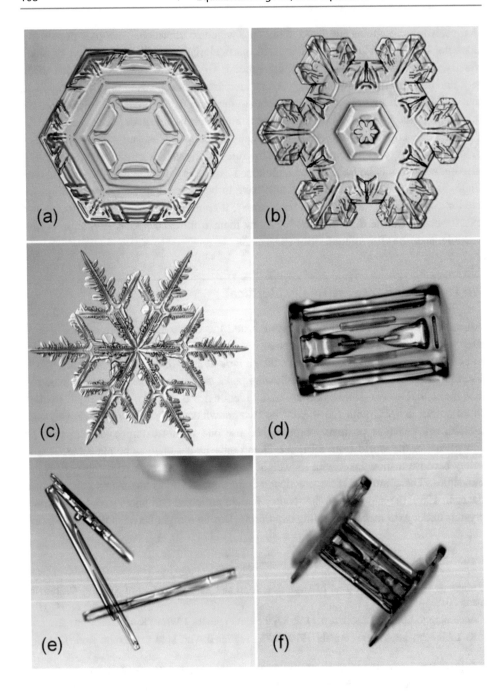

layers, she had caught the crystals on a glass plate coated with oil, which was attached to a rod mounted on the outside.[26] Now, she claimed that they were "identical", but that is also not true: They are very similar, but not the same, and not built like typical snow-flakes.[27] This cannot be cited as a refutation of Bentley's thesis.[28] So, let's continue to assume that no two snowflakes are identical, and if they were, they would be *extremely* rare exceptions Also: How would one find them?

Here, we see chance and coincidence for the first time in connection with our efforts to understand the causes of their endlessly widespread occurrence: Did all of the snow-flakes that we have just look at come from the same cloud? It is approximately just as cold everywhere, the humidity approximately just as high, and when the snowflakes fall down, everything looks quite uniform. Yes, but only *approximately* and only *quite uni-form*! Is a cloud in thermodynamic equilibrium? Certainly not! The weather is driven by the sun; there are fine small differences, fine small deviations in the conditions every-where. A first tiny water crystal experiences a number of changing environmental condi-tions as it tumbles down, while a flake is forming. Each original crystal, each growing flake follows a different path with slightly different conditions. So, it is unpredictable at which corner, at which edge, and how fast the next crystal will join the growing snow crystal. Kenneth Libbrecht has researched these processes and the results, and has basi-cally made it his hobby. This led him to publishing of a *Field Guide*.[29] You can think of it as a reference book, like those for the worlds of birds, flowers or mushrooms that you take with you on your hikes, for convenient use out in the field when there are snow-flakes to be found.

In his scientific publications,[30] he correctly points out that these are non-equilibrium processes that are responsible for the fact that no one snowflake looks like a second one—at least, not as far as we humans have been able to determine so far. They are typ-ical frozen dissipative structures. Kenneth Libbrecht is actually an astronomer and, as such, has been involved in the search for gravitational waves as one of the researchers in the large successful LIGO experiment of recent years.

And by the way: The world's first photos of snowflakes under a light microscope were taken in 1879 and 1881 by the Schleswig-Holstein lawyer and amateur naturalist Dr.

---

[26] https://journals.ametsoc.org/view/journals/bams/69/5/1520-0477-69_5_496.xml

[27] https://opensky.ucar.edu/islandora/object/imagegallery%3A2586

[28] https://opensky.ucar.edu/islandora/object/imagegallery%3A2586

[29] https://www.amazon.de/Field-Guide-Snowflakes-Kenneth-Libbrecht/dp/0760349428, see also his website http://www.snowcrystals.com/

[30] https://www.researchgate.net/publication/336715889_A_Quantitative_Physical_Model_of_the_Snow_Crystal_Morphology_Diagram

Johann Heinrich Ludwig Flögel from Ahrensburg, the neighbouring town of my place of residence. This fact was not known until 2010.[31]

A Japanese research group led by Takao Kameda from the *Hokkaido Snow and Ice Research Laboratory* published a classification of snowflakes and similar *solid precipitation* in 2013, based on observations from mid-latitudes to the Arctic and Antarctic polar regions.[32] They defined 121 categories of snowflakes that are fundamentally different in structure, up from 80.

Even under the conditions of a calm frosty night on the ground—approximately constant in comparison to the turbulent conditions during the formation of a snowflake in the air—ice flowers form in shapes that are similar to each other but never identical, and also not symmetrical. (Photo by the author in the Duvenstedt Brook nature reserve in Hamburg, Christmas 2021).

We move from the cold into the heat, albeit only partially. We stand at the geyser *Strokkur* in southern Iceland, where it is cool or even cold, and wait for the next eruption of hot water. Wikipedia says: " It typically erupts every 6–10 minutes."[33] Sometimes, one can also read, it *regularly* erupts every 6–10 minutes or is "periodically spouting".[34]

---

[31] https://www.shz.de/regionales/schleswig-holstein/panorama/wem-gehoert-die-erste-foto-flocke-id2018881.html, *all available in German only*, see also https://www.shz.de/lokales/stormarner-tageblatt/floegel-jahre-in-stormarn-id18903591.html*and*   https://de.wikipedia.org/wiki/Johann_Heinrich_Ludwig_Fl%C3%B6gel

[32] T. Kameda, K. Kikuchi, K. Higuchi, A. Yamashita, „A global classification of snow crystals …", Atmospheric Research 132–133 (2013), 460–472; the publication contains photos of all different categories.

[33] https://en.wikipedia.org/wiki/Strokkur

[34] https://en.wikipedia.org/wiki/Geysir

I was there many years ago with my family and checked it with my then-young sons, we measured and noted the waiting times. The geyser erupts much more irregularly. And quite often, it can be several hours until an eruption can be observed again, as we learned when we were there.

The cause is easy to understand: After a water outburst, water must first collect in an underground reservoir. It flows in from the groundwater. This afterflow cannot be 100% even, after all, it is not a well-functioning public water supply with uniform water pressure. While the water is flowing back and the quite narrow exit channel above the reservoir is also filling with water, it is heated by the hot rock under which the magma lies still deeper in the ground. This heating will also not take place 100% evenly, because sometimes the reservoir is almost empty, sometimes the geyser has spewed out less than before. Sometimes the rock is very hot, sometimes a little cooler, and it takes longer. The higher the water in the column above the reservoir—the water column can be very different in height—the higher the pressure in the reservoir. The higher the pressure there, the higher the boiling point of the water, well above 100 °C, and here, too, there are a few more or fewer (tenths) degrees. Now, steam bubbles rise up, some of the water from the overhanging column is pressed out, the pressure in the reservoir decreases—and suddenly, we have an explosive eruption, because the water is now overheated, explosively changes into steam and presses the overhanging water out with considerable speed. The geyser is applauded by the spectators. The natural scientist is pleased and sees a non-equilibrium system in action: An open system that exchanges matter—water—and a supercritical amount of energy—heat from the magma—with its environment and releases entropy in the form of water vapor, fine droplets and a slight increase in temperature in the cool ambient air of Iceland.

Our bodies and all other organisms are such open systems, and they are dynamic systems. William Bateson, creator of the term *genetics,* said: "In general, we think of animals and plants as matter, but actually they are systems through which matter constantly flows."[35] Merlin Sheldrake writes: "A mycelial network [of a fungus] ... reminds us that all forms of life are actually not *things,* but *processes.*"[36] This is the characteristic of non-equilibrium processes, which is why structures arise in them and through them.

The non-equilibrium system Earth contains an energy reserve from its formation billions of years ago, a reserve that, inter alia, is responsible for continental drift. Plate tectonics builds the mountains, but erosion and gravity together tear them down again, thus providing nutrients to lower layers and to the sea, to the ecosystems. In the middle ages of the Earth, as was recently reported after years of research, there was a particularly thin Earth crust about 2.5 to 0.5 billion years ago.[37] Therefore, hardly any new mountains

---

[35] Quoted from Merlin Sheldrake, Entangled Life, German edition "Verwobenes Leben", Ullstein 2020, p. 85.

[36] Merlin Sheldrake, loc. cit., p. 85.

[37] https://www.livescience.com/earth-mountains-disappear-boring-billion.html

were formed. The ecosystems became impoverished in terms of nutrients, evolution stagnated and came to a standstill for a billion years. When the Earth's crust became thicker again, plate tectonics was able to build mountains again, and so, about 300 million years ago, the *Cambrian explosion* occurred: Practically all known animal species originated in this period. May I recommend this to you as a sort of *learning module*? So, the Earth is in a non-equilibrium state. Life, evolution are non-equilibrium processes; they require the flow of solar energy and nutrients to export entropy. If continental drift slows down, the Earth is locally on the way to equilibrium, hardly any further changes take place, and evolution stagnates.

Now, let's also look at two oaks of the same age standing side by side.[38] Are they identical? No, they are not at all. Their size, the trunk diameter, the shape, where the branches branch off, how many there are, how thick they are, the bark pattern, everything is different in detail. Where the next branch and the next branch after that forms is left to chance and coincidence. We will very soon come to understand the causes of the occurrence of chance and coincidence in the next but one chapter.

If we jump to much larger scales, we find that no galaxy looks exactly like another; not only with snowflakes, we do not find two that are identical. In *Monthly Notices of the Royal Astronomical Society,* one could recently learn about the extremely variable structure and properties of galaxies:[39] One can conclude: Every galaxy is unique. The cosmic bodies appear with different bulges, halos, discs and rings. In some, many new stars are born, others hardly develop at all. There are at least 2×10^{18} galaxies (2 trillion),[40] and

---

[38] https://naturfotografen-forum.de/data/o/95/476656/image::dr.karl-heinz_limmer.jpg   With kind permission from Dr. Limmer.

[39] https://academic.oup.com/mnras/article/505/1/991/6123881

[40] https://www.nasa.gov/feature/goddard/2016/hubble-reveals-observable-universe-contains-10-times-more-galaxies-than-previously-thought

each one observed so far is like no other! There would be plenty of chance for at least two of them to be identical, but no …

Non-equilibrium is always dynamic, highly interesting, and often very beautiful. When we look at photos (like the one at the end of this chapter) of two galaxies in interaction taken by the *Hubble* space telescope, or the latest photo of the cirrus nebula (in the preface), we see an incredibly beautiful universe. Consider all the different animals, the birds and their feathers, best seen under a microscope! Or coral reefs that I could marvel at as a diver, or the complexity of a river delta like that of the Lena on this book's cover—our non-equilibrium world is beautiful! Now, let's imagine instead what an equilibrium world would look like: All substances—if they ever existed, because how could they have been generated if not in extremely-far-from-equilibrium stars—would be evenly distributed in space. There would be no suns that could produce heavy elements. No planetary systems with suitable planets that could enable the emergence of life. There would be no DNA containing a blueprint of organisms, no proteins that could act as enzymes, no feathers with which birds could hover in the air. There would be no us. Everything would be dead, but in equilibrium. That's not at all what we want. So, let's stop complaining that something has gone out of equilibrium. Fortunately for us, it was never in equilibrium to begin with, because, otherwise, we would not exist.

The photo below[41] *shows two interacting galaxies. There is nothing else like it in the universe. Like any other dissipative structure, these galaxies are unique in their structure, even though the patterns often resemble each other.*

---

[41] https://esahubble.org/images/heic1107a/Es it is of the object "Arp 273", 300 million light years away from us.

Almost Despairing of Science

5

Abstract

Surprisingly, non-equilibrium thermodynamics is hardly present at universities, rarely in research, and almost not at all in the curriculum, let alone elsewhere in society. This is despite the fact that the justification for awarding the Nobel Prize in 1977 to Ilya Prigogine clearly sets out how important this theory is for the fundamental understanding of our world. Most revolutionary new findings of science had a similar fate: it required a lot of time before they were widely accepted. But Albert Einstein's much more complicated theory of relativity was quickly recognized and has since been at the center of many popular science articles and books. Not at all so for Prigogine's theory. Thomas Kuhn has presented an explanation for the different acceptance of new ideas in his book *The Structure of Scientific Revolutions:* the *paradigm* based on which scientists are doing their work.

Ilya Prigogine received the Nobel Prize in 1977 "for his contributions to non-equilibrium thermodynamics, particularly the theory of dissipative structures".[1] Nevertheless, this actually quite important branch of thermodynamics still leads a *shadow existence.* And that, despite the fact that the world we live in consists essentially of non-equilibrium systems. This is also explained by the Nobel Prize Committee in its press release,[2] in which it says: "Classical thermodynamics has played a dominant role in the development of modern science and technology. It suffers, however, from certain limitations, as it cannot be used for the study of irreversible processes, but only for reversible processes and transitions between different states of equilibrium. Many of the most important and interesting pro-

---

[1] https://www.nobelprize.org/prizes/chemistry/1977/summary/

[2] https://www.nobelprize.org/prizes/chemistry/1977/press-release/

B. Wessling, *What a Coincidence!*, https://doi.org/10.1007/978-3-658-40671-4_5

cesses in Nature are irreversible. A good example is provided by living organisms which consume chemical energy in the form of nutrients, perform work and excrete waste, as well as give off heat to the surroundings without themselves undergoing changes; [...] Dissipative structures display two types of behaviour: close to equilibrium their order tends to be destroyed, but far from equilibrium, order can be maintained and new structures be formed. The probability for order to arise from disorder is infinitesimal according to the laws of chance. The formation of ordered, dissipative systems demonstrates, however, that it is possible to create order from disorder" [if and only if far from equilibrium, author's note].

I would like to point out a very important aspect in the the Nobel Prize Committee's justification here: Non-equilibrium processes are *irreversible* processes. This does not mean that one could not reverse certain processes with a lot of effort back to the starting point, but, if it is at all possible, only with *very* high effort. If we have made mayonnaise, we cannot recover the starting materials individually and put them back into the bottles. At least, we would certainly fail to bring the raw eggs back into their shells with the recovered egg yolk. We will not be able to return the egg yolk to its bladder, we cannot repair the shells, let alone the inner membrane of the egg. Please imagine that a sea eagle has flown over your house. I have one that does so occasionally. Could it be conceivable that it could regress back into an egg? The feathers grow back into down, the bones become soft, all organs develop back ... and suddenly there is the egg with a hard, unbroken shell back in the nest? No, that is impossible, just as it is impossible for a foal to regress back into the stallion's sperm and the mare's egg cell. I am not asking if time could suddenly reverse direction (that is not possible, a topic that we will discuss in Chap. 7). No, here, I am addressing the question of whether, during the course of the normally progressing time, a development that has once begun, a non-equilibrium process like fertilization, the development of the embryo and the birth, the subsequent growth and maturation, could reverse itself at some point. No, that is not possible; these processes are, in principle, irreversible.

We cannot make a boiled egg raw again by taking away the heat we put into it earlier, nor in any other way. The proteins have been irreversibly converted into another structure. We cannot remove the conductive polymer that we synthesized and then dispersed in a paint base. If we tried to filter it out with an ultrafine filter, we would separate the fine particles—in theory, it is practically impossible—but each nanoparticle would still be surrounded by a monomolecular layer of the matrix, the polymer basis, of which the paint base consists. How can we remove it and bring it back into the paint pot? It just doesn't work, not even with solvents! I tried this once, because I wanted to examine my product after dispersion; I found out: It doesn't work (see Appendix 8).

For what reason are these processes irreversible? It is the entropy: By initiating a process—a dispersion, the boiling of an egg, the fusion of a cell with sperm—entropy has been produced, ultimately exported into space, and we can not, under any circumstances, collect it again and bring it back into the process. This makes non-equilibrium processes irreversible. It looks different in equilibrium or near equilibrium: These processes are reversible. We can dissolve sugar in water and get the sugar back unharmed if we allow the water to evaporate. It happens by itself; it just takes a while.

Non-equilibrium also not Popular at Universities

Before I studied chemistry, Prigogine's theory was already finished, apart from the fact that such a theory is never complete. Shortly before I finished my Ph.D., he received the Nobel Prize, but I didn't take notice of it at the time. His theory was not taught at all during our studies, not even during the advanced studies. Now, I admit that thermodynamics anyway was not one of my favorite subjects; for me, it was rather a tedious subject that was compulsory to obtain my diploma. But ten years after I left university, I started dealing with it again. Because I was looking for explanations for the occurrence of these very complex structures and processes that I had discovered (Chap. 3). According to everything that could be read about *dispersion* and *colloid chemistry* or *physics* in scientific literature and textbooks, they were not possible and had not been observed before. Something like that is not even considered possible in relatively new textbooks.[3] So, it is no wonder that there is also no understanding of this outside the core scientific circles. This can be seen, for example, in the German-language edition of Wikipedia, but also in a seminar paper from the Max Planck Institute for Colloid Research.[4, 5] These are in line with generally accepted opinions, as shown in other references cited in[5]. Even in a book

---

[3] K. S. Birdi, Surface and Colloid Chemistry, CRC Press 2010, cf. http://bit.ly/36S34rd.

[4] Because many people like to save time by looking something up in Wikipedia and consider what they read to be reliable, I consider it necessary to correct the statements made there: (https://en.wikipedia.org/wiki/Dispersion_(chemistry)) "Dispersions do not display any structure; i.e., the particles (or in case of emulsions: droplets) dispersed in the liquid or solid matrix (the *dispersion medium*) are assumed to be statistically distributed. Therefore, for dispersions, usually percolation theory is assumed to appropriately describe their properties." (At least, if you read carefully, you will notice that the authors write "are *assumed* to be statistically distributed". They therefore claim not that it *is so*, but only that it is *assumed*. Now, we know more, and it is not as had been assumed: dispersions are, in fact, structured, a fact that has been experimentally proven. The *percolation theory* mentioned there is a theory that—even if doesn't say so—is based on equilibrium preconditions. The lack of understanding is also evident in another Wikipedia article: https://en.wikipedia.org/wiki/Dispersion_stability, because it is not the case that dispersions are *unstable* for thermodynamic reasons, as is stated there. Non-equilibrium does *not* mean *instability*.

[5] That the previous footnote is not the reproduction of a position that has long been revised in science, but is actually a widely held, yet nevertheless incorrect assumption, can be seen in this seminar given by the Max Planck Institute for Colloid Research (German): https://www.mpikg.mpg.de/886743/Emulsions_-2.pdf. Apart from the fact that scientific publications about fundamentally oriented colloidal research are also scarce in the English language, what can be found describes the same basic understanding: the dispersed or emulsified phases are homogeneously statistically evenly distributed within the medium, cf. T. G. Mason et al., "Nanoemulsions: formation, structure, and physical properties", J. Phys.: Condens. Matter 18 (2006) R635–R666, https://www.researchgate.net/publication/230981868_Nanoemulsions_Formation_structure_and_physical_properties, especially Fig 1 on page R638, which is definitely a wrong description; cf. also the graph titled "Examples of a stable and of an unstable colloidal dispersion" in https://en.wikipedia.org/wiki/Colloid.

with the title *Structured Fluids,* which I bought many years later and which deals with numerous beautiful examples of different structures, non-equilibrium thermodynamics is not mentioned, nor the term *dissipative structures*. Instead, it is based on *statistical thermodynamics,* a branch of *classical* thermodynamics that deals with states and processes *in the neighbourhood* of equilibrium.[6]

When I realized, at the end of the 1980s, that the structures I had found could be understood and explained with Prigogine's non-equilibrium thermodynamics, I looked for a corresponding chapter in my thermodynamics textbook from my university days. I didn't find one. So, I bought a new revised edition of my old *Atkins*.[7] You will find a little bit of relevant information in the book's nine hundred pages, but, in the end, only twenty pages deal with this so very crucial field of thermodynamics. Not much more can be found in the latest edition of this most important and most widespread textbook of thermodynamics for chemistry students.[8] So, it is not surprising that the following idea is still widespread in science: Everything—or at least most of it—can allegedly be explained with the concept of *equilibrium,* at least, with the assumption that these states or processes are close to equilibrium. *Non-equilibrium,* states and processes far from equilibrium, are thought not to exist at all or, if they do, then they are treated as mere exceptions, not worth mentioning. I hope I was able to convince you that there is hardly a more inappropriate basic assumption for the description of nature, our world and the entire universe.

Prigogine and his theory, or the principle of *dissipative structures,* have indeed obtained the recognition of the Nobel Prize Committee and, of course, of a certain number of scientists as well, but unlike Einstein's rather more complicated theory of relativity, Prigogine's theory is largely unknown. But this is or was the case mainly in *the West*. In the then-GDR and -Soviet Union, Prigogine's concepts fell on fertile ground very early and found a much broader scientific basis. The same applies to Haken's theory of synergetics. This was essentially due to sociological reasons. The *self-organization,* which Prigogine described on the basis of a new thermodynamics, was also interpreted by the scientists as a sign of hope, as a light at the end of the tunnel in the GDR and USSR, both of which were urgently in need of reform. The *Perestroika* in the USSR and the subsequent upheaval in 1989 in the GDR were therefore interpreted as a *phase transition* completely in line with Prigogine's and Haken's theories.[9]

Watson and Crick's revolutionary findings practically immediately found general recognition. Einstein's theory of relativity was not immediately accepted and only prevailed after hard discussion and in the face of fierce—but mostly fair and highly qualified, abso-

---

[6] Thomas Witten, Structured Fluids (Polymers, Colloids, Surfactants), Oxford 2010.

[7] Peter W. Atkins, Physikalische Chemie, VCH Weinheim 1988 (German), my original edition is 20 years older.

[8] Peter W. Atkins and Julio de Paula, Physikalische Chemie, Wiley VCH 2010 (German), see http://bit.ly/3rvj9uV.

[9] R. Feistel, W. Ebeling, "Physics of Self-Organization and Evolution", Wiley-VCH 2011, Preface.

lutely typical and necessary—resistance, and then ultimately relatively quickly, within only a few years.[10] Both revolutionary new findings were and are still taught, at least in advanced courses, in high school. There are numerous easily understandable—of course, simplifying!—books about the theory of relativity, and one does not have to be an X-ray structural analysis expert to understand that DNA has the structure of a double helix. When Professor Röntgen discovered the "X-rays", it did not take long before "X-raying" became really popular and was advertised in widely sold newspapers. In public squares in the cities and at fairs, you could have yourself X-rayed. That this was not recognized as unhealthy at the time is another matter.

Scientific Breakthroughs and Inventions Ususally have a Hard Time

Such a quick—let alone a broad—recognition of scientific achievements and breakthroughs is anything but normal. Scientific revolutions usually only prevail after a long time. Max Planck once said: "A new scientific truth does not prevail in such a way that its opponents are convinced and declare themselves to have been instructed, but rather because its opponents gradually die out and because the new generation is already familiar with the truth."[11] We do not have to look back to Copernicus and Galileo, who were put under terrible pressure for their outstanding and revolutionary scientific achievements from today's point of view. The theory of continental drift, for example, was fiercely opposed and rejected during the lifetime of the polar explorer Alfred Wegener. Only a few decades after his death in the Greenland ice was his theory generally accepted. The tenacity of established views is enormous. Resistance from circles of the scientific establishment is the norm. A quick acceptance of scientific or technical breakthroughs—that is, completely new scientific findings or inventions that are later celebrated as "breakthroughs"—is much rarer.

But please, let's check ourselves: How do we imagine the continental drift, the plate tectonics? I imagined it as a drifting apart of the great former *primordial continent Pangaea*. But according to latest studies, our picture of it looks different now, given that it was shown how a chaotic coming together and drifting apart of different plates was involved over one billion years.[12]

---

[10] A good summary of the discussion can be found at https://www.mpiwg-berlin.mpg.de/news/features/features-feature7, another article also highlights the tough discussions in the early years after Einstein's publication: https://daily.jstor.org/why-no-one-believed-einstein/

[11] Max Planck: Scientific Autobiography, cf. https://philpapers.org/rec/PLASAA-3, cited and translated after the German edition: Johann Ambrosius Barth Verlag, Leipzig, 1948, p. 22

[12] https://www.businessinsider.com/video-earth-tectonic-plates-billion-years-2021-2?r=DE&IR=T

In an issue of *Spektrum der Wissenschaft,* a number of such breakthroughs have been described that have prevailed in the end despite massive resistance from reviewers.[13] Among others, the first working laser: The inventor of the laser (in this case, a *Maser*) was rejected several times by a physical journal, but then accepted by the journal *Nature.* The evolutionary biologist Lynn Margulis had it a little harder: her work was rejected by 15 scientific journals. A year later, she managed to convince a biological journal to publish her research results—since then, the view has gradually prevailed that cell organelles (e.g., mitochondria) are the result of the symbiotic fusion between bacteria and evolutionary predecessors of today's cells, which were organelle-free. However, the theory she had proven had already been put forward as a hypothesis decades earlier, which gives you an idea how much time it took. Gregor Mendel also had a very hard time, he had to fight for years to get the results of his pea experiments published. They were only recognized much later as the *Mendelian inheritance rules.*

In areas completely outside of natural sciences, new theories that can explain the contradictions of new observations with regard to established theories are rarely accepted quickly, too. This is also what the psychologist Daniel Kahneman experienced with his and his colleague Amos Tversky's *Prospect Theory.*[14] I have occasionally touched on their psychological foundations in this book because they tell us something about human thinking. He notes that his theory, which was awarded the Nobel Prize, is by no means reflected in the textbooks of macroeconomics.[15] A situation that is almost identical to that of Ilya Prigogine's non-equilibrium thermodynamics. Kahneman adds a rather friendly, sympathetic explanation for this, saying: Even the simplified, if unrealistic, basics of macroeconomical science would be difficult for students to understand; they would first have to learn about the tools, and only later could the *more difficult toolbox* be used. Recently, a professor I know made a similar statement about his thermodynamics lectures when I asked how far he had gone in taking non-equilibrium theory into account: He could not expect this of the students as long as they did not even understand equilibrium thermodynamics. "Later", it would be possible. But if there is practically no room in the curriculum, and only twenty (!) of the nine hundred pages in the standard textbook are reserved for this subject, when is *later* ? When do students of the natural sciences come into contact with the more complicated toolbox of non-equilibrium thermodynamics? I am afraid what this means is—never (and I experience this in conversations with young chemists and physicists): Because this way, their view of the natural sciences and how the world is functioning was completely shaped by the concept of equilibrium.

Such experiences are difficult to digest for researchers, who are convinced that they have recognized something fundamentally new and want to communicate it to the sci-

---

[13] https://www.spektrum.de/wissen/publikationen-die-dem-widerstand-der-gutachter-trotzten/1308502

[14] https://en.wikipedia.org/wiki/Prospect_theory

[15] D. Kahneman, loc. cit. p. 352.

entific community. However, we cannot even imagine how Ignaz Semmelweis felt. He had recognized that the lack of disinfection of the doctors' hands was the most common cause of childbed fever. He was laughed at and attacked, and his findings were discredited as speculative nonsense, although he had systematically proven it. He could not get his life back on track, and was later arbitrarily—without diagnosis—locked up in a psychiatric institution. He died there shortly afterwards under dubious circumstances. Today, many years after his death, he is famous, his story serves as material for films and books, and his findings have become common knowledge. Robert Anton Wilson coined the term *Semmelweis reflex*.[16] This is understood to be the *reflexive* rejection of a new scientific discovery by established scientists without any objectively unbiased review. The discoverer is fought instead of supported, one refuses to even discuss the new findings openly and fairly, because the new view contradicts established, deeply rooted convictions and undermines the reputation of the leading scientists in the field.

Professor Harald zur Hausen was also rejected and laughed at when he published his new findings on the origin of cervical cancer, which (at the time) was the second most common cancer in women. He had found, in 1976, that it had to be a papilloma virus, but could not prove this directly at first, because the test methods available at that time could not detect the virus in *sleeping* tumor cells. Not until 1984 did he and his working group succeed in proving its presence in tumor cells. Previously, towards the end of the 1970s, cytological screening[17] began to be established very slowly in Germany, having started in the USA. It became a common occurrence for someone to present zur Hausen's new findings in congresses of the time, a fact that regularly led to heated arguments, even to shouting, as the prevailing view considered herpes viruses to be the cause.[18] It was not until 2006, i.e., 18 years after the final proof, that a vaccine was approved. Harald zur Hausen received the Nobel Prize in 2008.[19]

mRNA research, the basis for the Corona vaccines from BioNTech and Moderna, is a prime example of both our central question, *How do chance and coincidence enter our world?*, as well as the question that undergirds this chapter, *Why is non-equilibrium not understood, but instead rejected?* It took less than a year from the identification of the SARS-Cov-2 virus to the release of the first vaccine. In total, there are at least five decisive coincidences that made this development occur so quickly, as far as we can tell from the outside. Probably, there are still further countless small ones that have served the pro-

---

[16] https://en.wikipedia.org/wiki/Semmelweis_reflex

[17] Microscopic examination of cell swabs.

[18] Personal communication from my life partner, who was one of the first cytology experts in Germany and who was trained, theoretically and practically, at these congresses.

[19] https://www.nobelprize.org/prizes/medicine/2008/press-release/ see also https://www.welt.de/wissenschaft/article2537376/Der-Mann-der-ein-Medizin-Dogma-zerstoerte.html (article in German, translation of the article title: "The man who destroyed a medicine dogma").

gress of research from time to time, because, since the end of the 1980s, hundreds of scientists have contributed to mRNA science and technology:[20, 21]

Coincidence 1: Biotechnologist Dr. Katalin Karikó continues to research therapeutic uses for mRNA, even though she was practically demoted to a postdoc position at her university in the USA. This is because she was unable to raise research funds, allegedly proof that this line of research was pointless. Many years after her initial discovery that mRNA could possibly be used to treat diseases, she discovered, together with a colleague, how to circumvent the innate cell defense: The genetic code of synthetic mRNA had to be slightly modified. The discovery remained unnoticed for many years. The decisive factor for the later success was also that the technology for embedding mRNA in nanoscopic lipid particles was developed in the laboratory of Peter Cullies in the second half of the 1990s. This enabled the mRNA to enter the cell unharmed.

Coincidence 2: In Tübingen, Ingmar Hoerr is working on his doctorate; his goal is to find out whether an immune reaction occurs in mice when they are injected with DNA. As a control experiment, from which he expects nothing but a *null reaction,* he uses the corresponding mRNA copy. To his surprise, the mouse control group showed the strongest immune reaction![22] A few years later, he founded, together with a number of other people, the Biotech company CureVac. The startup initially focused on cancer therapy, and later also on lung diseases, Lassa fever, yellow fever and rabies. In 2017, an mRNA-based vaccine failed because there was no increased survival rate compared to a placebo group. He later reported from a conference: "In the front row, a Nobel Prize winner stood up and shouted: 'This is complete bullshit, what you're telling us—complete bullshit!'"

Coincidence 3: In the USA, a researcher recognizes the possibilities and founds Moderna. In Germany, Uğur Şahin and his wife Özlem Türeci grasp the potential of this still fledgling technology a few years before Moderna. It is completely unclear whether it can be made usable. The two German oncologists want to develop cancer therapies based on this and found BioNTech.

Coincidence 4: The brothers Andreas and Thomas Strüngmann, who have become billionaires through the sale of the company Hexal, which they had founded and operated, decide to trust the Şahin/Türeci couple and invest in BioNTech as early as 2008.

---

[20] Cf. https://www.spiegel.de/wirtschaft/unternehmen/biontech-gruender-oezlem-tuereci-und-ugur-sahin-deutschland-wird-genug-impfstoff-bekommen-a-00000000-0002-0001-0000-000174691195 as well as https://en.wikipedia.org/wiki/RNA_vaccine; https://en.wikipedia.org/wiki/Katalin_Karik%C3%B3;https://www.focus.de/politik/ausland/katalin-kariko-im-teddybaer-schmuggelte-sie-ihr-startkapital-ungarin-ist-mutter-der-corona-impfstoffe_id_12754185.html; https://www.welt.de/wissenschaft/plus221314612/Covid-Impfstoff-Warum-der-mRNA-Impfstoff-das-perfekte-Mittel-gegen-das-Coronavirus-ist.html

[21] https://www.spektrum.de/magazin/der-verschlungene-weg-zum-rna-impfstoff/1974538, article title translated: "The labyrinthine path to the RNA vaccine", it also features more details of the mRNA story (German).

[22] Interview with Ingmar Hoerr in ZEIT on May 5, 2021, see https://www.zeit.de/2021/19/cure-vac-mrna-impfung-corona-ingmar-hoerr-franz-werner-haas. (German)

The founder of SAP, Dietmar Hopp, invests in CureVac, even at a time when the company doesn't seem to offer anything other than opportunities. According to Hoerr, a lot of luck and networking played a role. He told the German news magazine SPIEGEL: "It's a coincidence that Curevac and BioNTech still exist, namely because private investors financed them even though there was nothing to sell."[23]

Coincidence 5: In early 2020, when the Corona pandemic has not yet even been classified as a pandemic, BioNTech changes course and puts all its eggs in the *Corona vaccine* basket. The company has not yet had a cancer therapy approved, nor has the practical proof of the medical suitability of the mRNA technology for oncology been provided, but the researchers use their knowledge and infrastructure to develop a Corona vaccine as quickly as possible. And the EMA approves it only 10 months after work had begun.

So, the originally rejected research approach by Dr. Karikó was able to become successful many years after her demotion and the surprising, unexpected discovery by Dr. Hoerr after just as many years. For Şahin/Türeci's company, which was originally focused on cancer therapy, these were years of research in the no-man's-land of knowledge and of constantly fragile and precarious company-building on the brink. Dr. Karikó is now working in a leading position at BioNTech since many years.

It is remarkable to look at how mRNA vaccines are stabilized so that they can be made available and effective in the body: They are incorporated into extremely small lipid particles. Obviously, this is the much more difficult and time-consuming part of production, both in terms of raw material procurement and storage/transportation: In order to prevent them from settling, that is, from clumping, from *agglomerating*, they must be stored and transported refrigerated. That's colloid chemistry! What a coincidence, then, that mRNA research and stabilization began long before the Corona epidemic struck the world!

Even though lipid stabilization was probably researched and developed *empirically*, it is nonetheless part of colloid research, i.e., the science of dispersions and emulsions with nanoscopically small, finely dispersed or emulsified particles. It is the type of mixture that I recognized as not being an equilibrium system. Conducting more scientific work and, at the same time, also providing broader education at universities and to the public, to contribute to the understanding of *non-equilibrium* and to engage in fundamental research, all that has been overdue for many decades in all areas of the natural sciences.

It is incomprehensible why, even after the award of the Nobel Prize to Prigogine in this area, there was no upturn at universities or in the broader public; there is not a single popular science book about this important theory. In stark contrast to the anything but intuitive theory of relativity, or quantum physics or, more recently, black holes. I would like to encourage you to think about what the causes could be. To do so, I will share a few insights from my own treasure trove of researcher's experience, which contains

---

[23] https://www.spiegel.de/wirtschaft/unternehmen/curevac-gruender-ingmar-hoerr-ueber-risikokap-ital-und-die-konkurrenz-zu-biontech-a-aa45e040-9b4a-4b6e-9382-6c79459da83a

some findings on the subject of dispersions and emulsions, systems that Prigogine and his colleagues did not address. The trove also contains some experience in dealing with a scientific community that is confronted with new, revolutionary findings.

Dispersion is Underestimated

As you will remember, I started by trying to disperse conductive polymers. I explained the conceptual difficulties to you: If these substances consisted of long chains arranged in the finest long fibrils, as was generally claimed, *dispersion* would have been the completely wrong approach. Most of the researchers in this huge[24] international research field focused on trying to *dissolve* these conductive polymers. Because, for various basic studies, and if one wants to put any possible application of these new materials into practice later on, one has to be able to process them, i.e., bring them into the desired form. But at first, this was not possible at all: melting, dissolving, such methods did not work. That's why I came up with the idea of dispersing them. But, for that, I was not allowed to tear up the long chains in the fibrils, that was the general opinion and my fear …

Fortunately, the conductive polymers showed me in my studies that they consist of very small, tiny spheres. How the chains are arranged therein, where the electrons move along—I didn't worry about that at first. I simply assumed first that these new polymers would not be soluble in common solvents, and especially not in thermoplastic plastics and lacquer binders. That was the complete opposite of the general opinion! And I could not know whether my initial assumption was correct.

But, in the end, we were finally able to disperse them. We had developed a special synthesis for our preferred conductive polymer (Polyaniline) that made it dispersible,[25] and we made a very surprising discovery: If we carefully dispersed a special form of our polyaniline in a solvent and slowly increased the concentration, we suddenly obtained a dry rubber-like mass at concentrations of only 1 to 2%—but 98.5% of this *rubber* consisted of solvent! If you touched this mass, your fingers stayed dry! At some point, one of my laboratory assistants dropped such a piece: It bounced back like a *superball (bouncy ball)*.[26] We made our jokes about how we should go into the toy market with the conductive polymer, instead of into the high-tech market. But we would probably have failed because of our expensive raw material. At that time, I examined our polyaniline-*bouncy ball* (after freeze-drying it) in the electron microscope and found extremely

---

[24] At the peak in this research field, probably more than 20,000 researchers worldwide were active at the same time, in universities, research institutes and all the leading big companies of the chemical and electronics industry.

[25] Cf. Appendix 6, Part 2.

[26] https://en.wikipedia.org/wiki/Super_Ball

complex pearl chain structures.[27] Why was the material dry? Because all solvent molecules were adsorbed on the gigantic surface of the only 5–10 nm small nanoparticles that had formed complex three-dimensional chain structures. A clear proof: dispersions are highly structured, the nanoparticles are not homogeneously, not statistically evenly distributed, but they form complex three-dimensional networks, as was later found for mayonnaise, yogurt and many other dispersions and emulsions (see Chap. 3). And soon after, we achieved the first increasingly conductive samples that we generated by dispersion in plastics or lacquers and that I could show at congresses.

While we were thus achieving ever better research results, I was increasingly eyed with suspicion by the large worldwide scientific community that was researching conductive polymers. They had no confidence in our results and tried to slow me down. How could such a small company with so few people and so little money achieve such allegedly good results? That couldn't be true. Disbelief on the one hand and envy on the other dominated. At that time, I did not yet know that envy is the highest form of recognition. The Chinese say: 宁被人妒, 不受人怜。(Nìng bèi rén dù, bù shòu rén lián): Better to be envied than pitied. But I was unable to view the situation around me in that way at that time. I openly admit: The lack of acceptance made me feel insecure and tormented me enormously.

The then-director of the Max Planck Institute for Polymer Research, with whom I worked before and after the first publication and whom I occasionally met for discussions, said to me, after one of my lectures at a conference, in a fatherly, but, at the same time, arrogant tone: "Oh, you know, Doctor Wessling, I think it would be better for you and the company if you only took care of the quality of your products. That will occupy you more than enough. Your approach to the subject is misleading." With that, our cooperation ended. I didn't even answer him, I just turned around and left. A few years later, when I already had a commercial breakthrough based on my research, he said to some of his employees, from whom I heard about it: "Come on, this Wessling just wants to talk up his successes! There is nothing behind it."

Despite such resistance, I kept trying to promote an understanding of the principle of dispersion at congresses and seminars. I showed, with new data and facts time and again, that conductive polymers were not soluble, but could be dispersed, and that one could not tell a true solution from a colloidal dispersion with the naked eye. This is not even possible with a light microscope, because the particles are much smaller than the wavelength of the visible light. Solutions and dispersions are both transparent and can be coloured. Polyaniline colloids are transparent and deeply green. I called it *maygreen,*

---

[27] B. Wessling, "Dispersion hypothesis and non-equilibrium thermodynamics: key elements for a materials science of conductive polymers", Synth. Met. 45 (2), 119–149 (Nov. 1991), especially pp. 134 and 135; available here: https://www.researchgate.net/publication/233979584_Dispersion_hypothesis_and_non-equilibrium_thermodynamics_key_elements_for_a_material_science_of_conductive_polymers_A_key_to_understanding_polymer_blends_or_other_multiphase_polymer_systems.

because it is a shade of green that is quite similar to the fresh leaves in May. PEDOT and polyacetylene dispersions are blue. But nobody was interested, even though these were simply verifiable facts. A few years later, I also showed theoretically, using a thermodynamic approach, that they are, in principle, insoluble.[28]

Every two years, we researchers in the field of electrically conductive polymers and related materials held a large conference, the *International Conference on Science and Technology of Synthetic Metals(ICSM)*. Out of the more than twenty thousand researchers from universities, research institutes and industry worldwide, about two thousand scientists participated each time. I was the only non-professor on the International Advisory Committee of about thirty scientists. In 1992 in Gothenburg, I gave another lecture on this subject of "conductive polymer dispersion" at this conference. Professor Alan Heeger, the later Nobel laureate and one of the scientific community's most powerful opinion leaders, wanted to make me look ridiculous during the discussion minutes after my presentation, saying: "Bernhard, these are only words, 'solution' or 'dispersion', what you are doing is only playing with words." He received applause amidst general laughter, and probably assumed that he had silenced me with it. The audience thought that my nonsense about the insolubility that could be circumvented by dispersion had finally been debunked: all just words, it was only wordplay that I was engaging in. My spontaneous answer was: "Alan, you are right, only words, like 'plus and minus', 'electron and positron', 'proton, photon and neutron', all that is the same, only words." My reply was quite a sarcastic one. Suddenly, it was as quiet as a mouse. Hundreds of listeners, who had just laughed loudly and slapped their thighs, did not move, did not cough and did not clear their throats. Because, of course, everyone immediately remembered that terms are strictly defined in science in order to be able to distinguish things and phenomena from each other and to enable communication. They represent an agreement. I had deliberately chosen terms from physics for my spontaneous answer because Heeger is a physicist.

Of course, Alan Heeger would know that *proton* and *neutron* are also only words, but that they denote something quite different. Something that can be distinguished by measurements. That the terms I cited as examples actually denote opposite phenomena, he knew, but that *solution* and *dispersion* are also opposites, apparently not: Solutions are thermodynamically considered equilibrium systems, dispersions are non-equilibrium systems—that's how simple it is in principle. In solutions, the molecules are completely isolated and surrounded by solvent molecules, solvated. In dispersions/emulsions, only the outer surface of the particles or droplets, which contain thousands or millions of molecules, is wetted with the molecules of the dispersion medium, with a monomolecularly

---

[28] https://www.researchgate.net/publication/202290125_Conductive_Polymer_Solvent_Systems_ Solutions_or_Dispersions; this text later appeared in a chapter I conitrbuted to a handbook, as subchapter 2 of the whole chapter: https://www.researchgate.net/publication/253651172_Conductive_Polymers_as_Organic_Nanometals.

thin layer. Solutions have no internal interfaces, dispersions have extremely large internal interfaces, which is what makes them so different and enabled our *bouncy ball*. Solutions are Newtonian fluids, dispersions/emulsions are non-Newtonian.[29] There is therefore nothing that solutions and dispersions have in common, except that they, for example, have water or organic solvents as main components if they are liquid dispersions (there are also solid ones). And they can be transparent, and possibly colored for the eyes in most cases if the dispersed particles are significantly smaller than 300 nm. In every other respect, solutions and dispersions are totally different, they are opposites.

I explained this briefly to the participants and then left the stage. Most of them probably did not understand it, they had never heard of *non-equilibrium* before, and if so, they considered this term synonymous with *unstable,* that is, something bad, unwanted (see also Appendix 11, Part 2).

Led Astray by Outward Appearance

The problem with this question has been and still is that, for most physicists as well as for many chemists, *solution* and *dispersion* mean approximately the same thing, which, however, is far from being the case. They are actually stark opposites. How deeply rooted the conceptual confusion is, and how little clear it is even among chemists, including colloid chemists, that these are *opposites*, is shown by the confusion of terms in the designation *colloidal solution,* a term that occurs everywhere including in the colloid science literature: What is this supposed to be, a colloid, i.e., a dispersion, or a solution, which would thus not be a dispersion? How can a system be both a dispersion (a colloid) and a solution at the same time: "colloidal solution"? For me, it's like calling a horse a *female stallion.* Horse lovers would experience that like a screeching sound in their ears. A *square wheel* or a *spherical hexahedron* are just as impossible as the *quadrature of the circle,* all of that is considered to be an *oxymoron* everywhere.[30]

Many scientists from this research scene have judged and classified solutions and dispersions according to their appearance, with or without the use of a light microscope. By this, they mean that everything that is transparent and clear and stable is a solution, while something that is rather turbid and unstable is a dispersion. But that is not the case! Because a stable dispersion or emulsion with a dispersed or emulsified phase that has a diameter of about 100 nanometers or less is also transparent, because our eyes can only resolve particles in a light microscope that are at least as large as the wavelength of visible light, i.e., at least 300 nanometers. One of the most famous and oldest colloids (dispersions) is *Faraday's Gold,* it has been stable for over 150 years and is transparent ruby

---

[29] https://en.wikipedia.org/wiki/Newtonian_fluid or https://en.wikipedia.org/wiki/Non-Newtonian_fluid.

[30] https://en.wikipedia.org/wiki/Oxymoron

red in color.[31] Dispersions containing particles with about one micrometer particle size and more tend to be turbid and unstable.

If I had studied Chinese proverbs back then, I would have rubbed the doubters' noses in the following: 按图索骥 (Àn tú suǒ jì); or, freely translated into English: "Judging a horse only by pictures":[32]

> Once upon a time, in the "Age of the Spring and Autumn Annals" (春秋时期 chūn qiū shí qí, a period of Chinese history from 722-481 BC), there lived a man named Sun Yang (孙阳 sūn yáng). He was an expert in the appraisal of horses and had even written a book about it. People called him Bo Le, after the celestial being responsible for the heavenly horses.

> One day, his son decided to go out and find a great horse as well. However, he had never really seen a horse before, but relied on a book. He was excited to have found an animal with a prominent forehead, two big eyes, and four big hooves. "This must be an outstanding horse!" he thought. Proudly, he brought the animal home and showed it to his father. The boy said, "Father, I found a very good horse, except that the hooves are not really good."

> Sun Yang was unsure whether to laugh or cry, because his son had brought home a toad. Sun Yang said to him, "My son, this horse is not bad, but it loves to jump too much, and you can't ride it."

Educated Chinese use this Cheng Yu when they want to criticize people who do not think, but thoughtlessly and superficially follow a picture, in order to judge something, instead of trying to understand the nature of a thing, a phenomenon.

So, I suffered for years from enormous self-doubt. It was a tightrope walk of emotions from "I'm about to make myself look ridiculous" to "My findings are correct, stay calm". Was I a charlatan or was I doing serious science? Were my results just speculation and self-deception or were they solidly interpreted and scientifically sound? The path between the dark abyss of "quack and impostor", and thus "stubborn, incompetent chemist who doesn't realise that he's completely wrong", on the left and the sunny summit of "we're bringing polyaniline dispersions to the market" on the right was barely visible. I stumbled across unexplored, extremely narrow and by no means stable ground. I constantly felt like I was about to slip into the depths—because I didn't know for years how it would turn out and what the scientifically correct finding really was. In addition, there was the big question of whether we would ever be able to become commercially successful with the results. Again and again, depressive moods lasting days or weeks afflicted me, partly because of repeated exclusion within the scientific community, partly aggravated by acute financial problems. The liquidity was always in short supply. I didn't yet know Kahneman's term "sunk costs", which one should no longer take into account when thinking about abandoning a probably hopeless project or continuing it;

---

[31] https://www.rigb.org/our-history/iconic-objects/iconic-objects-list/faraday-gold-colloids

[32] https://blogs.transparent.com/chinese/5-more-awesome-chinese-idioms/, text changes by the author.

he advocates for quicker project cancellations.[33] But I certainly knew the German saying that would be translated into English as "don't throw good money after bad". After all, I was not, as in Kahneman's examples, just a manager who had fallen in love with his project—which, of course, I was in love with—and whose self-esteem and career opportunities were at risk if the project was cancelled—that didn't matter to me at all. On the contrary: I would have risen many notches higher in the esteem of my co-shareholders, who also had strong doubts about the prospects of this expensive research project, if I had "abandoned the project in view of the increasing cost risks". There was no fundamental conflict of interest between me and the other shareholders, because I was myself a big shareholder and, of course, interested in only pursuing long-term successful and marketable projects. After all, I had to pay off a bank loan worth millions that I had secured to buy the company shares. So, these were heavy inner battles I was fighting.

But then, one day, I found Thomas Kuhn's book *The Structure of Scientific Revolutions.*[34] In it, he explained to me, without addressing me directly and without knowing that I would become his reader, that I should not be surprised that my findings were not accepted, not understood, indeed, *not even ignored* by the scientific community. The reason for this is that almost all scientists work on the basis of a *paradigm*, on the basis of a common starting (world- or research field-)view.

The Power of the Paradigm

Most scientists, according to Kuhn, work on tiny sections of the scientific edifice, which stands on the foundation of a paradigm.[35] As long as the observations of the scientists working on the very small aspects of the respective area do not—at least not obviously—contradict the expectations based on the paradigm, everything is fine. Initial contradictions are ignored, further contradictions are explained with ever more adventurous, ever more complicated theoretical treatises, as can be seen in the epicycle theory to explain the partially apparently backward planetary motion in the Ptolemaic world system.[36,37] Only when the majority of scientists have become intolerably dissatisfied with an

---

[33] Daniel Kahneman, *Schnelles Denken, langsames Denken,* loc. cit p. 423 ff. (original English edition: *Thinking, Fast and Slow*), see also https://en.wikipedia.org/wiki/Sunk_cost

[34] Thomas Kuhn, "Die Struktur wissenschaftlicher Revolutionen", German edition Suhrkamp 1996; original edition: "The Structure of Scientific Revolutions", University of Chicago Press 1962.

[35] Examples of old outmoded paradigms: The earth is a disc. Or: The sun and the stars revolve around the earth. Or from the beginning of this chapter: Cervical cancer is caused by herpes viruses, so it can't be papilloma viruses.

[36] https://en.wikipedia.org/wiki/Deferent_and_epicycle

[37] https://en.wikipedia.org/wiki/Geocentric_model

ever-growing lack of agreement among observations, measurements and expectations based on the generally accepted paradigm, are they willing to allow a revolutionary replacement of the previous one with a new paradigm. This was the case with Einstein and his theory of relativity within a few years, not least because his theory was able to explain the previously completely incomprehensible movement of the planet Mercury. Certainly, the proof of the deflection of the light of a star by the gravitational field of the sun by means of measurements on photographic plates, obtained during a solar eclipse, convinced the previously doubting physicists even more.

The worldwide approximately twenty thousand researchers working in the conductive polymers field obviously felt no discomfort at all about inexplicable phenomena in the late 1980s/early 1990s. They saw no discrepancy among theory, prediction and experimental observations. In my opinion, this was because these scientists carried out and evaluated their experiments exactly as they had seen in various previous publications. They did not know the brutal test of industrial practice that I subjected my products to. In an article in the German weekly newspaper *DIE ZEIT,* I once said to Klaus Jopp, a science journalist, during his research on my company Ormecon for a longer article: "We can pour our hypotheses into buckets, so to speak."[38] This meant: hypotheses that can be implemented reproducibly into products on a large scale are no longer hypotheses, but scientific facts, confirmed hypotheses, theories that work in practice. The other 19,999 researchers did not know this practice, so they did not notice any contradictions to and within their theory.[39]

But I was in the midst of practice: My research results should flow directly into products and then into the market. So, an unyielding corrective judged me: not the academic and practice-distant peer-reviewers of a scientific journal (only if I wanted to publish there), but the customers we sampled or supplied with our products! The products had to have reproducible effects. We had to be able to produce them reproducibly on a much larger scale than just once in a tiny amount of a few milligrams or a gram. It had to be possible on the scale of tons, and that for customers worldwide. We worked on that, and we achieved it. Reproducibility of results is actually the essential criterion for scientifically acceptable results. And, in contrast to university researchers, my success criterion was not the number of publications or the award of a prize, not the winning of a competition for research funds, but rather: The products have to work; they have to do what we promise or what the customers demand from them; they have to be produced reliably and

---

[38] https://www.zeit.de/2000/43/Gezaehmte_Polymere/komplettansicht

[39] Kahneman's book "Thinking, Fast and Slow" provides an explanation from another perspective here, which he calls "theory-induced blindness": In his field of research, too, those scientists who represented the established theory did not change their minds even when they were confronted with blatant contradictions in observations related to their theory (loc. cit., P. 340 ff.). J. Kounios and M. Beeman (Chap. 3, Ref. 45) formulate it as follows: "Everything we think has the potential to limit what we think next." (loc. cit. German edition p. 44) We think in "boxes" that are difficult to get out of.

bought by enough customers so that we can finance further research with the proceeds, and so the previous research investments can pay off.

The Power of Images

I understand the term *paradigm* in the meantime in a sense that extends further than the one presented by Kuhn. A paradigm is more than a generally accepted scientific basis. It acts—if it is still an unproven hypothesis—in some cases, like a *dogma*. One *believes* in a dogma. According to my understanding, a paradigm, a dogma, does not even have to be formulated in writing, but preferably digs itself into the heads of scientists in form of a picture. These are very powerful pictorial images. Therefore, even scientific terms such as *dispersion* can be understood differently, because they create different images in the heads of different scientists: For me, particles with an adsorbed shell in complexly structured pearl necklaces flash up as images. For the vast majority of scientists, however, a picture with particles or molecules evenly distributed in a medium would show them immediately emerging—because *solution* and *dispersion* were considered to be approximately the same. A *conductive polymer* is associated with the mental image of fibrils and stretched chains along which the electrons hop. Against this, one does not first come with scientific arguments or measurement results, even showing that the apparent fibrils consist of extremely tiny globules does not change anything. Because giving up a (hypothetical) research paradigm is also equivalent to the collapse of one's own scientific world view. After all, one would have to admit to oneself or—even worse—to outsiders that one has worked in vain for years or decades in the wrong direction.

But what do people who want an economy in equilibrium or the restoration of *ecological equilibrium* have for an image of *equilibrium*? To answer this question, it is worth looking into school books. As an example, we look at the online version of the school

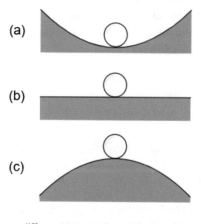

different types of equilibrium

physics book *Physik Libre*.[40] All three different positions of the ball in the above Figure from the online physics book are referred to as being in *equilibrium,* namely, as stable (a), indifferent (b) and labile (c) equilibrium. But in the end, isn't *labile equilibrium* an oxymoron, that is, a conceptual contradiction in terms, just like *dynamic equilibrium, Fliessgleichgewicht?*

What do scientists' mental pictures of *equilibrium* look like? They are basically not much different from those in the school book mentioned above (see the graphic below left): A ball falls or rolls down from a height, from a state with a lot of potential energy, into equilibrium, i. e., into a minimum of energy. In addition, there are other equilibrium states with similarly low energy. These can be reached due to a slight disturbance. Non-equilibrium, on the other hand, does not exist in the world of imagination of most scientists. At best, they imagine *labile* or *dynamic equilibria,* terms from a world view that is still shaped by the equilibrium paradigm. However, the image of non-equilibrium could look like the graphic below: In this way, one can schematically describe a non-equilibrium process that leads to a *comparatively stable non-equilibrium state over a certain period of time.*

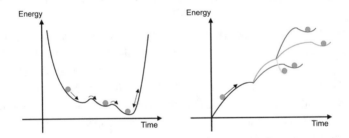

Left graphic: Equilibrium; right graphic: Non-equilibrium.[41] The y-axis shows the energy of the system, the x-axis the time. In non-equilibrium, the system repeatedly experiences situations (forced by the supercritical amount of energy flowing into the system) in which different paths can be taken: Bifurcations that lead to different relatively stable states (symbolized by different colors) via different paths, each of which occupies a much higher energy level than an equilibrium state.

---

[40] https://physikbuch.schule/equilibrum-and-stability.html, Michael Grundel, *Physik Libre,* Freely available physics book for high school level II, graph taken from https://commons.wikimedia.org/wiki/File:Types-of-stability.svg.

[41] Graphic concept adopted (but graphically changed and clarified) from B. Wessling, https://www.researchgate.net/publication/233979584_Dispersion_hypothesis_and_non-equilibrium_thermodynamics_key_elements_for_a_material_science_of_conductive_polymers_A_key_to_understanding_polymer_blends_or_other_multiphase_polymer_systems.

You can imagine this by thinking of throwing a ball with a lot of energy from the street in front of a house onto the roof, watching it roll down the roof tiles and settle in the gutter. You can also imagine a high mountain lake without drainage: it rains, the lake fills up, the water stays in the mountain lake and does not flow into equilibrium, i. e., into the open sea. In addition, over time (x-axis), while energy is constantly flowing through the system, other states suddenly become possible. We will think about this in the next chapter. A non-equilibrium process does not necessarily lead to exactly *one* result, but can cause several or even many different results. These are then to be practically understood in the description of the *energy landscape,* as high mountain valleys in the heights of the potential energy that the respective system has reached.

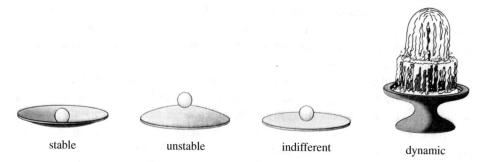

stable   unstable   indifferent   dynamic

In a book's chapter on *Ecology,* the term *ecological equilibrium* is introduced and often used as if it were self-evident, and it is in general use.[42] First, the term *equilibrium* is illustrated with an image from a physics book (see the figure above). The text then explains that the *labile equlibrium* does not adequately describe biological systems, but that a picture of a flowing fountain (upper right) is more appropriate: dynamic equilibrium.

This is similar to the term Fliessgleichgewicht *(flow equlibrium),* which we discussed in Chap. 4. Then, the authors continue: "When something flows, it does so because it is *not* in equilibrium." How true! And furthermore: "Every biological system uses a complex *texture of imbalances* … " (emphasis by the authors). And yet, the whole thing is entitled *The ecological equilibrium,* and the cycles of the most important nutrients in a pine or oak forest and the phosphate cycle are schematically represented and explained. Both *cycles* seem to move on their own (just like the fountain in the picture, whose

---

[42] e.g. here: E. Philipp, B. Verbeek, Ökologie, in: Materialien für den Sekundarbereich II, Biologie (= *Ecology,* in: *Materials for Secondary Level II, Biology);* Schroedel-Verlag 1998, p. 72; reprint permission by Westermann-Verlag.

pump, which cannot run without energy from outside, is not visible), because the schemas do not show the crucially important solar energy, without which nothing moves. Despite the clear emphasis on *non-equilibrium ("imbalances")*, the pattern of *ecological eqilibrium* and *cycle* (which looks as if it can work on its own) remains as a learning goal. This paradigm is imprinted as an image in the minds of the students.

Amazing and, at the same time, characteristic of how widespread the concept of equilibrium is, how powerful the images of equilibrium are, is this: Prof. Dirk Brockmann, physicist and chairholder with a research focus on "Complex Systems" at Humboldt University in Berlin, has written a very remarkable book about complexity in our world. He also writes about ecosystems in it. According to him, they would be in a stable equilibrium or should develop back into one. And, on pg. 138, he shows the concept of equilibrium in the form of a graphic exactly as I have depicted it above left: balls roll into an energy minimum, which corresponds to a maximum entropy; that means "equilibrium".[43] Complexity (minimum entropy) such as that in an ecosystem cannot arise in equilibrium (maximum entropy).

The images of the *macroeconomic equilibrium* that we briefly mentioned in Chap. 4 are more abstract. This *equilibrium* is dealt with in a textbook on the basics of economics.[44] Here, the graphics suggest a linear behavior of the economy—at least, they describe this as desirable, as it is also stated in the German Constitution—in which supply and demand for goods, money and labor are to be "in equilibrium". However, we all experience every day that the economy behaves extraordinarily non-linearly in contrast to this. Here, in the pictorial world of human beings, scales or see-saws are probably more likely to play a role, as in the graphic below from an English-language finance dictionary, not so much balls that roll into a depression.[45] While, as worded there, "general equilibrium" is in balance in the entire economy, this is only the case in individual sectors of the economy in "partial equilibrium" (in the following graphic to the right).

---

[43] Dirk Brockmann, "Im Wald vor lauter Bäumen—unsere komplexe Welt besser verstehen", dtv 2021, p. 135 and 138. (German)

[44] Dennis Paschke, *Grundlagen der Volkswirtschaftslehre – anschaulich dargestellt*, PD-Verlag 2011, Chap. 10; if anyone should think that this is perhaps just one man's opinion, I refer them to the fact that the 5th revised edition (2007) alone sold 27,000 copies; the book is used in introductory lectures in economics and in economic high schools, which speaks to its wide acceptance and large readership.

[45] https://marketbusinessnews.com/financial-glossary/general-equilibrium-definition-meaning/

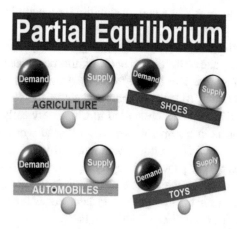

We see that *equilibrium* is deeply and widely anchored in our language usage and in our pictorial imagination. It appears lightning fast and intuitively, and apparently explains everything to us.[46] *Imbalance* and *non-equilibrium* are not terms with positive connotations for us. We also do not associate it with a familiar mental image. Even if we were to neutrally use the equally appropriate term *dynamic systems* which can be stable or unstable, no new image would arise. A simple term, like *flow* or *flux*, used instead of *dynamic equilibrium* (and instead of *steady state*), would be appropriate, exemplary and at the same time image-generating. Because, ultimately, all organisms, the ecosystems, our economy, even our solar system, the Milky Way and the universe are processes, hence *flows*—energy, matter and, not least, entropy *flow*. Already, looking at *consumption,supply* and *demand*—none of these are ever static numbers or states, but *flows (fluxes)*: Here, the goods flow from the producers to the consumers, just as how the raw materials that we get from the earth flow to the producers. The money flows back, the entropy flows into the garbage dumps and, as waste heat, into the environment, and ultimately into the universe. So, everything always changes, it flows. We change our environment actively, like all organisms, and like the weather and the climate. But even

---

[46] According to Daniel Kahneman ("Fast Thinking, Slow Thinking"), our thinking system 1 provides such images and prevents system 2 from dealing more intensively with the more complicated relationships (which is very energy-consuming; real thinking is exhausting!); see footnote 3 in the foreword. According to Kahneman (loc. cit. p. 114), the consistency sufficient for System 1 is decisive. "It is easier to fit everything you know into a coherent pattern if you know little." Most people know very little or nothing about the non-equilibrium properties of our world; culturally, linguistically and pictorially deeply anchored equilibrium concepts provide coherent images and apparently sufficient explanations.

if everything changes, there is still a certain stability (non-equilibrium is not necessarily unstable, on the contrary!).[47]

In China, the knowledge of constant change was already deeply rooted in philosophy thousands of years ago: The oldest book of the classical Chinese philosophical texts is the 易经 (yì jīng, the *The Book of Changes*), which is said to date its origins back to the third millennium BC.[48] It is characteristic of the philosophical view of the world that emerged and became deeply rooted in China very early on, when the Bronze Age was beginning for us in Europe, a view according to which everything is in constant change.[49] For the Greek Heraclitus fifteen hundred years later, "nothing is so constant as change"—πάντα ῥεῖ panta rhei, everything flows.[50] Nevertheless, especially in our Western cultural sphere, the images of *equilibrium* and *durability* have become particularly ingrained. In regard to what would later be so fondly called *dynamic* (or: *flow) equilibrium (Fliessgleichgewicht)* by Bertalanffy, as if everything remained the same despite the flow, Heraclitus said 2500 years earlier: "Into the same rivers, we both step and do not step; we both are and are not." And this too: "No man ever steps into the same river twice." These could have been memorable opinion- and culture-forming images, but in our cultural environment, the images of equilibrium and constancy have been and are much more powerful. *Flow equilibrium (steady state, dynamic equilibrium)* and *(dynamic) flow (flux)* are not interchangeable terms, if only because 50 to 60% of all rivers (with water flow, or: flux) worldwide dry up at least one day a year, even in the wet regions of Southeast Asia, where 35% do so.[51] From this, we see that even our natural rivers *(water fluxes)* are non-equilibrium systems and by no means can serve as examples of Bertalanffy's *flow equilibrium, or Fliessgleichgewicht (dynamic equilibrium)*, let alone *steady state.* For not even these are stable in *equilibrium,* what's more: They do not meet Bertalanffy's definition requirements for *flow equilibria.* Only those systems in

---

[47] An alternative term to *dynamic equilibrium (flow equilibrium = Fliessgleichgewicht)* that correctly captures the non-equilibrium property but does not use the word 'equilibrium' was requested of me by one of the readers of an earlier version of the manuscript when I rejected previously accepted terms. He considered it appropriate to use the term 'homeostasis'. But apart from the fact that that term is probably not known to most people outside the profession, it again suggests *equilibrium,* that is, from my point of view, it is just as inappropriate, because the origin of the term (https://en.wikipedia.org/wiki/Homeostasis) lies in Ancient Greek for "similar" and "standing still", which is not how dynamic systems behave; and it is understood as "self-sustaining equilibrium"—which is also not what self-regulating systems do; because the relatively stable dynamic state that dynamic systems (rivers) reach is only possible through external inflow of energy and exchange of matter (import and export).

[48] https://en.wikipedia.org/wiki/I_Ching

[49] https://en.wikipedia.org/wiki/Chinese_philosophy

[50] https://en.wikipedia.org/wiki/Heraclitus

[51] Mathis Loic Messager et al, Global prevalence of non-perennial rivers and streams, nature Vol. 594, 17 June 2021, p. 391–397, see https://www.nature.com/articles/s41586-021-03565-5

which the macroscopic variables, here, the water levels, do not change are called as such. But that changes all the time! A term like *(dynamic) flow (or: flux)* on the other hand, describes the properties of such non-equilibrium systems in an appropriate way.

The power of images is also reflected in our languages. Those who shaped the predecessors of the German language in the past seem to have been intuitively clear: We think first in images.[52] Why else is it called (in German) Welt*bild,* An*sicht,* Sicht*weise,* Blick*winkel,* Meinungs*bild,* if it is about complex opinions or positions in words and writing? Even the English *view*point hearkens back to an image when it comes to opinions. And also in Chinese, we *see* the same root of the translation of *Meinung/Ansicht (view* point): 观点 guān diǎn, practically a literal translation of *viewpoint,* with us in German *Sicht*weise.

In my research, I thought a lot using mental images and videos. Often, I made drawings of what I wanted to understand. This was, for example, what happened in the chaotic airport of Newark, when I was drafting the model for the step sequence of how the sudden conductivity breakthrough at the critical volume concentration happens. Also the qualitative basis of my non-equilibrium theory and the turbulence theory of non-Newtonian liquids[53] originated first in mental images and videos, in my thought experiments.

The Chinese have a saying for this as well: "Seeing is a hundred times better than hearing": 百闻不如一见。(Bǎi wén bù rú yí jiàn.) This quote was the answer given by the general Zhào Chōngguó (赵充国) to the question by the Han emperor Xuāndì (宣帝, 91-49), as to how many soldiers he needed to repel an invasion of the Huns (匈奴 Xiōngnú).

Let's look at another proverb that supports my thesis of pictorial thinking: "A picture is worth a thousand words." 一画胜千言 (Yī huà shèng qiān yán.) We all know how powerful pictures are when used in the media—photos and moving television images that convey messages much more impressively than words. The ancient Chinese seem to have known this.

But surprisingly, this proverb is not of Chinese origin. It can even be dated to the day it was put into circulation, namely, December 8, 1921. On that day, Fred R. Barnard published an advertisement in an American advertising industry magazine with the slogan "One Look is Worth A Thousand Words."[54] The advertisement promoted pictures in

---

[52] I also wrote about "thinking in images" in my book "The Call of the Cranes" with considerations about how cranes (which I have studied in particular) or animals in general think, to the extent that they are mainly visually on the move—because the idea that animals think (ie, weigh decision options), is now generally accepted, and I have also concluded from my diverse observations and further research that cranes actually do think, actually do weigh options. For this purpose, I formed the hypothesis that they could do this in images (including dynamic ones).

[53] Together with Helmut Baumert, who worked out the mathematical-theoretical part in this case, while I contributed "only" the qualitative basis. You can find out more about this work in Appendix 13.

[54] https://en.wikipedia.org/wiki/A_picture_is_worth_a_thousand_words

advertising prints on trams. In 1927, a second advertisement appeared with "One Picture is Worth Ten Thousand Words." The number had miraculously increased to ten thousand within only six years.

Barnard later explained that he had passed the slogan off as a Chinese proverb so that people would take it seriously. The proverb spread quickly and was soon—as could be expected—attributed to the Chinese philosopher Confucius. Apparently, we trust the ancient Chinese with a lot of valuable wisdom and knowledge. A more in-depth discussion of the power of images in worldwide research on conductive polymers can be found in Appendix 14.

Retreat from the Research Area with a Bang—Did Anyone Hear It?

The power of the established paradigm was apparently unshakeable. So, I decided, after more than thirty years, to leave this society. That's why I gave a final talk at the regular large international conference *ICSM 2010* in Kyoto (Japan), at which I presented our latest research results.[55] I later explained these in detail in a scientific (and, as usual, peer-reviewed) publication (for more details see Appendix 12).[56] There, I described a result that actually should have led to great interest and further work in other laboratories. It is about a chemical reaction during the application of the Organic Metal in corrosion protection or in printed circuit board technology: During the *passivation*—the first and crucial step, when my Organic Metal comes into contact with iron or copper, see Appendixes 9 and 10—chemically precisely defined metal complexes are formed between this nanometal and iron (2+) or copper (1+) (depending on the substrate), recognizable by a completely changed UV-VIS spectrum: completely new and highly interesting compounds!

I also reported on a discovery that I had only recently made: The organic metal polyaniline has a helical basic structure; although the chains are very short (only about 12 monomer units), they form a short helical turn and, when the metallic state is forcibly achieved by dispersion, they coalesce into an effectively longer helix. Actually, there should have been great interest in these results, especially since I showed signs that, after complex formation, the Cu(1+) or Fe(2+) ions were located in the screw axis of the helix. All of this should have actually resulted at least in verbal contradiction, if not better in follow-up work for confirmation or refutation and to check how it looks with other representatives of this class of substances—but nothing of the sort happened. This was an extremely clear sign of the very regrettable decline of this area of research, which I

---

[55] https://www.researchgate.net/publication/237904229_B_Wessling_ICSM_Kyoto_2010-07

[56] B. Wessling "New insight into Organic Metal Polyaniline morphology and structure", Polymers 2010, 2 (4) 786–798, freely accessible here https://www.mdpi.com/2073-4360/2/4/786/htm.

had foreseen and described twenty years earlier.[57] There are only a few research groups active in this area today.

For me, the last lecture and the subsequent publication closed a circle—what an irony of non-equilibrium, chance and coincidence: As a teenager, I admired Watson and Crick, their discovery of the DNA double helix was one of the triggers that made me want to become a biochemist. I did not become one. "Sai Weng Shi Ma"—Sai Weng loses a horse, how can one know that it is not a blessing? I lost my way to biochemistry and instead found the way to a completely new substance that, by chance, also has a helical structure. I was allowed to discover it and even develop the new organic metal into useful, marketable products ... What a gift of coincidences in this wonderful life.

Let me end this chapter with a few personal thoughts before we trace the birth of chance and coincidence in non-equilibrium in the next one: My private life, my researcher life and my entrepreneur life are (like the lives of all people and all living beings) a mixture of non-equilibria. I have been allowed to experience an enormous number of positive and negative experiences and adventures, perhaps much more than the poor farmer Sai Weng in the story given at the beginning of the book. I will always be extremely grateful to Thomas Kuhn for his book *The Structure of Scientific Revolutions*; without him, I would probably have succumbed to my despair and given up. I have constantly been under pressure from all sides, have navigated through phases of liquidity crises, have collected successes and failures. As a young development manager, I was able to avoid bankruptcy twice, and as an entrepreneur, I myself narrowly avoided bankruptcy three times—in retrospect, I don't want to miss any of it. When the big crises were heading for their climax, of course, I couldn't and didn't want to enjoy them, because Sai Weng hadn't told me his story by then. When I washed up on China's shores, I was very frightened. I was afraid of this huge and incomprehensible country, with its completely different culture and the incomprehensible language and script. But coincidences threw me there, and I grew from it.[58]

不登高山, 不显平地。(Bù dēng gāoshān, bù xiǎn píngdì): Those who never climbed the high mountain do not know the flatlands.

不经冬寒, 不知春暖。(Bù jīng dōng hán, bù zhī chūn nuǎn): Those who have not experienced winter do not know the warmth of spring.

More details on contents in this chapter in Annex 6, 8 to 14.

---

[57] B. Wessling "Talking Point: Intrinsically conducting polymers: A critical stage in research strategy?" Adv. Mater. 3 (10) 507–509 (1991), to be found here: https://www.researchgate.net/publication/227978812_Talking_point_Intrinsically_conducting_polymers_A_critical_stage_in_research_strategy.

[58] B. Wessling, "Der Sprung ins kalte Wasser – in China mit offenen Augen und Ohren leben und arbeiten", Verlagsgruppe Eulenspiegel, scheduled for mid 2023. Translation of the book title: "Jumping in at the deep end – living and working in China with eyes and ears open".

The Birth of Chance and Coincidence in Complex Systems

6

Abstract

Using characteristic examples from biology (biochemistry, evolution), weather, climate change, complex networks and cosmology ("the big bang"), the dynamics and non-linearity of non-equilibrium systems are examined in more detail. It becomes clear that higher levels of organization of matter develop their own, new laws in comparison to lower levels. The *emergence* of new properties and laws is an important aspect. The *non-linear* behavior of these complex systems is the cause of the occurrence of chance and coincidence. Finally, the phenomenon of *decoherence* shows us why the quanta (elementary particles), with their indeterminacy, cannot cause chance or coincidence in the macroscopic world.

In the previous chapters, we learned about many examples of different coincidences, but we focused on the type of coincidence and chance that Monod had called *essential*. We left out those processes that only appear as coincidental to us because it is not possible or too expensive to predict the result for practical reasons; such processes that can also be reliably investigated using probability calculations should be referred to as *random*.

We have dealt somewhat extensively with the principle of *non-equilibrium*, which pervades and shapes our world. And you may have already suspected that the reason why I have talked about it so much, is because chance is somehow related to it. That's right, that is the case, and now, we want to get closer to the solution of the question *How do chance and coincidence enter our world?* so that we can finally answer it.

We had already learned an important criterion of non-equilibrium processes: irreversibility. I had described this using the example of the development of living beings from seeds and eggs, but I can also take a banal example from my favorite sport, soccer: If the ball has entered the goal (in accordance with the rules), then the goal is scored and will not be taken back, even if the ball only landed in the goal because of an unfortunate coincidental encounter with the foot of a defender who wanted to prevent a goal shot. This had changed the ball's trajectory so that it landed in the goal, but whereby, without the foot, it would have actually landed outside of it. Let's look at the flight path of the ball: After the striker kicked the ball, it followed a ballistically calculable path, in principle. For good mathematicians or physicists (although not for me), this is calculable from the acceleration and the spinning that the ball experiences, the angle that the path initially forms with the plane of the soccer field, the air resistance that slows down the ball, and the gravity that will eventually force the ball back to the surface of the earth.[1] During the game, even good physicists or mathematicians, if they are also good goalkeepers, would not be able to fall back on such a calculation. But the experienced goalkeeper can, even if he cannot handle formulas well on his laptop, *intuitively calculate* such balls, and knows where they will arrive. But now the defender's leg gets in the way: The beautiful linear movement, which is easy for the goalkeeper to assess, gets a kink in it, the ball continues to fly, although a little slower, but now, unfortunately, in a different direction, namely, into the goal. For a goalkeeper like me, that's extremely annoying, because the reaction then comes too late. For the affected team, a disaster, if this exact goal seals their relegation from the previous league.

This introduces us to another principle that is characteristic of non-equilibrium processes far from equilibrium: non-linearity. Every *event* is the result of a *non-linearity* in the previously linear process. This process exhibits fluctuations that can build up and make the process unstable, and suddenly, we have a new situation: Structures or processes have changed, an event has happened. Over the course of evolution, processes have been established for life that run extremely non-linearly: A chain of independent coincidental events is the cause of a surprise, such as the appearance of a new property in organisms as a result of mutations. A *new* property is the result of an initially linear process, in which (just as an example, a number sequence), after "1 - 2 - 3 - 4 - 5 -", not "6", but non-linearly "17" or "minus 31" follows. We observe this not only in evolution, but in life in general or in the course of the cells of every organism, as well as in inanimate nature and everywhere else in the universe. Such a thing is known as a *phase transition,* or also as an *event.*

Enzymes: Results of Non-Equilibria and Actors Therein

Enzymes are not simply constantly active. There are molecules that cause enzymes to start or stop their task. The function of the molecules controlling the enzymes is not linear. There are threshold values above which the effect reacts like a switch (ON or OFF).[2] This is done by changing the folding of the enzyme so that it can either work or not work.

---

[1] https://www.researchgate.net/publication/322335277_Study_of_soccer_ball_flight_trajectory.
[2] J. Monod, cit. loc. p. 72/73.

This, in turn, points us towards another very important characteristic of non-equilibrium systems: functions that are not contained in the individual parts suddenly arise when these parts come together. There are, for example, enzymes that are only active when four proteins come together, in many cases triggered by a starter molecule. They are then not really chemically (covalently) bound, but only adsorbed on each other; but then and only then is the enzyme active. The parts separate when corresponding trigger molecules make them do so. This is extremely non-linear, going, in one step, from zero to 100, and then back to zero.

Evolution has developed countless examples of this principle, but not in such a way that it would be too easy for us humans to decipher them. If you write down the amino acid chain of one of the four parts of such an enzyme, it looks as if it could fold three-dimensionally in almost unlimited numbers of arrangements, at least, from a purely statistical point of view. But the formation of the structure (the folding of the linear amino acid chain) into a complex three-dimensional object is dependent on an incredibly large number of factors. It is not statistics that would allow us to understand the folding, but the chemistry and physics of the interactions. They allow for only one energetically optimal arrangement, one that always arises quite spontaneously and of its own accord, without any external influence. While the sequence of amino acids is genetically determined, the three-dimensional structure cannot be read directly from the genetics. It arises exclusively from the interactions of the amino acid components with each other in the chain, and that spontaneously. No theory can be developed from even the best, detailed knowledge of the individual properties of the isolated amino acids that would predict the folding of a protein that is formed from these amino acids in a sequence determined by the DNA. Nor could the function be predicted. This is not fundamentally different from plastics (polymers) that are formed from monomers: The properties of polyethylene (a thermoplastic, i.e., a plastic that can be deformed by heat) cannot be derived from the properties of the monomer ethylene (a gas). Polymers—a substance group that includes proteins—are on a higher level of organization of matter than monomers.

With polymers, the properties *also change with the number* of monomers connected to each other, i.e., with the *degree of polymerization*. Let's look at the example of the CH_x groups in hydrocarbons. Methane, a gas, is CH_4, ethane, with 2 carbon atoms, is $H_3C\text{-}CH_3$, a gas as well. Six CH_x groups of these units is $H_3C\text{-}CH_2\text{-}CH_2\text{-}CH_2\text{-}CH_2\text{-}CH_3$, hexane, a liquid. If you have fifty to two hundred such units, it is a polyethylene wax, which is used, for example, in motor vehicles as a sealant to prevent corrosion in cavities. With several thousand such groups, it is polyethylene, which—depending on the microstructure and density—is used for packaging, pipes, cable and pipe insulation.

When you look at the thousands of known functional proteins in organisms and try to understand in which order the amino acids are chained together, there is only one rule: There is no rule. The order of the amino acids is purely random. In other words: If we knew all the amino acids from the beginning to the 99th, and then all the others from the 101st to the end of the chain, could we predict which one would be missing at number 100? No. Or: Let's take a few proteins with similar but slightly different functions. We could compare ones that bind water to double bonds of different substances. Could we predict the amino acid sequence of a fourth enzyme with a similar function from the amino acid sequences of three such enzymes? Again, no. So, it is random or by chance that the amino acids are arranged in a certain way and that certain functions are made possible by this. *By chance?* Yes, but in such a way by chance with material (biochemical, genetic) and researchable causes that we just can't predict the result, even if we know the DNA composition prescribing the amino acid sequence; at best, we can only somehow understand and explain it afterwards.

Until recently, it was not possible for researchers to derive the three-dimensional structure of an enzyme from the arrangement of amino acids in the protein chain after folding, to predict the structure of the enzyme. It is simply too complicated to know all the interactions with each other in space and to weigh them against each other. With the large number of interactions arise a large number of possible arrangements, which are energetically very similar. For decades, the only possibility was X-ray structure analysis, which is extremely difficult to evaluate for such large and complex molecules. That is why the structure of only a very small fraction of human enzymes is known. Only recently has it apparently become possible to predict the real structure based solely on the knowledge of the amino acid chain.[3] The company DeepMind (which belongs to the Google parent company Alphabet) developed a program that, with machine learning ("artificial intelligence"), was able to predict structures by learning from an enormous amount of known protein structures. However, the reliability is not yet sufficient: In test runs in which the computer was presented with unknown amino acid sequences, the structure of which has been clarified, the program was able to represent 70% correctly. Nevertheless, we should be aware that this is not based on a comprehensive physicochemical theory of protein folding, but on a trained AI that *does not understand* what it is doing.

While DNA therefore *only* contains the arrangement of amino acids as information, that is, a—if you will—purely linear information, a three-dimensional structure arises from it *of its own accord*. The information content of this structure of an enzyme is therefore much higher than that of the corresponding sequence in the DNA! A greater amount of information is equivalent to a more complex structure, and therefore lower entropy—this has been exported from the cell under energy input from outside the cell. This confirms to us once again that we are dealing with non-equilibrium. But there is something else that we really need to be aware of: At this level of aggregation of matter, which is higher than that of DNA, new properties suddenly appear that cannot be derived from the properties or information of the lower level.

---

[3] https://www.nature.com/articles/s41586-019-1923-7.epdf?author_access_token=Z_KaZKDqtKz-bE7Wd5HtwI9RgN0jAjWel9jnR3ZoTv0MCcgAwHMgRx9mvLjNQdB2TlQQaa7l420UCtGo8vYQ39gg8lFWR9mAZtvsN_1PrccXfIbc6e-tGSgazNL_XdtQzn1PHfy21qdcxV7Pw-k3htw%3D%3D.

Higher Aggregation Level of Matter: New Properties

We look at another level, into the cell: Here we find complex regulatory systems. Of course, these had not been created on purpose. Rather, out of the *possibilities* offered by the most diverse substances with their interactions, such regulatory systems have emerged in cells over the course of evolution, which finally gave rise to organisms capable of surviving and reproducing. From the *infinite* number of *possibilities*, those cell and organelle functions that are actually valuable or even indispensable for life and reproduction have emerged through *chance*. At the level of the cell, more new properties (= information!) have arisen that are not inherently contained in the single proteins and enzymes as such; they arise only by their interaction with each other.

For example, ribosomes are a product of self-organization, the spontaneous formation of order from disorder. Prigogine calls it *order from chaos:* Ribosomes are the location, the organelle, where proteins are synthesized. There, the translation of the code written on DNA and the amino acid sequence corresponding to the DNA code takes place. Ribosomes are the *protein factory* of the cells. Organisms with a true cell nucleus and a rich division of the cells, that is, subunits separated from each other by membranes, have very complexly built ribosomes. They consist of more than fifty different proteins and, depending on the organism, at least three ribosomal RNAs. If you separate the respective components and bring them together outside a cell, they spontaneously form fully functional ribosomes. The components do not form covalent, that is, real chemical bonds with each other. There are only electrostatic and adsorptive interactions that create a ribosome. Order from chaos, through physical interactions alone. And again, a new complex function arose on a higher level of organization of the substances. This is very much in line with Haken's theory of synergetics, which describes "processes in which the macroscopic state of the system changes qualitatively through self-organization (emergence of new qualities)".[4]

The amino acid sequence results from the DNA code. The DNA code "does not know" what three-dimensional structure results from it in the protein, let alone what function the protein will have. This results from the spontaneous self-organization due to the interactions. How the proteins are arranged together and the cell organelles formed results, again, only from the interactions that can be derived neither directly nor indirectly from the DNA or the amino acid sequence. The cells interact and form organs and bodies. These then form eco- and social systems. From the lower aggregation level *DNA,* we can there-

---

[4]Already >50 years ago, physicist P. W. Anderson published a groundbreaking article in Science (Vol 177, S. 393–396, 1972) entitled "More is Different". In it, he makes it clear that complexity arises with the appearance of more components in a system and thus new properties ("emergence"). He therefore does not believe it to be appropriate to state that one could understand all phenomena and laws on more complex levels of matter by means of an ever more detailed analysis of the lower levels of their components. See also section "Self-organization" in Chap. 4

fore neither directly nor indirectly derive the properties and behavior of human societies from the intermediate aggregation levels of protein, cell, organism, species, ecosystem. On each higher level of organization of the substances, new interactions occurring there are causing new functions. New laws arise that cannot be reduced to the single components of the respective level. Chance and coincidence shaped life, and continue to do so.

We must keep in mind: The fundamental structure of all organisms is the same. The control of all cell processes is based on the same mixture of substances, DNA and RNA, as well as proteins and other substances, which consist of only a few building blocks, and always the same ones. The components arranged in the DNA and RNA chains, in turn, contain the same purine and pyrimidine bases, each of which also fulfills the same type of function. In the end, it is only variations in detail that distinguish an amoeba from a sea eagle or us humans. But what huge differences in properties these variations cause!

I recently came in for a big surprise[5]: Evolution has developed an actually unimaginable ability in octopus and squid species that no other family of species has at its disposal. They can vary RNA, so that they can adapt by implementing instructions from DNA differently, namely, more advantageously than is normally the case in all other organisms through *copy/paste*. Here, an editing function is also active! 60% of the RNA in the nervous system is modified in this way, which also changes brain physiology itself. This seems to allow for adaptation to different water temperatures and, so it is assumed, is somehow responsible for the high intelligence of this species group. What new, surprising properties arise from the different interactions of the substances DNA/RNA/proteins known to us that allow a brain to be formed!

Let's now look at much simpler systems than the cells of organisms or evolution. Let's look into our everyday lives: Even a simple traffic jam "due to high traffic volume" can not be understood from the properties of the components that form the jam. A traffic jam does not arise on the level of one *car* or even a lower level, like *motor, wheel, transmission*. Not even on the level at which we describe the limbs and other organs of humans (including their brains), who drive the cars or sit in them. We cannot find or develop any regularities for the formation of traffic jams on a sunny Saturday morning on these lower levels of the elements involved in the traffic jam. We cannot even unify the laws that prevail on the elementary levels, which can describe a traffic jam formation. Of course, a traffic jam would not be possible without many people who steer their cars on the same road at about the same time, and not without these cars having engines and wheels and much more. But nevertheless, the laws of traffic congestion build-up cannot be reduced to the laws of the functioning of cars or drivers—and not even of both together. The laws of jam formation arise only when the traffic participants come together in large numbers and influence each other, that is, *interact* with each other and with the environment, the road network—something quite typical of self-organization processes.

---

[5] https://www.sciencealert.com/octopus-and-squid-evolution-is-weirder-than-we-could-have-ever-imagined

Just as the principles of evolution cannot be deduced from the properties and functions of plants, animals, and fungi themselves or, even less so, individual cells, let alone the individual components in the cells, the functions in a cell cannot be reduced to the properties of the thousands of individual substances that are involved, or deduced from the properties of these compounds. And, of course, the structure of these partly extremely complex molecules cannot, as such, be derived from the properties of carbon, hydrogen, oxygen, nitrogen and sulfur, nor the salts or metal ions that are also often involved, let alone from the respective properties of the protons, neutrons and electrons that make up these elements.

The non-equilibrium thermodynamicists Werner Ebeling and Rainer Feistel have expressed this understanding in their book *Chaos and Cosmos—Principles of Evolution* with words that we all probably know or have already heard: "Complex systems have emergent properties, *the whole is greater than the sum of its parts.*"[6] Here, we encounter the term *emergence* again.[7] It will appear again later. By *emergence* we mean the origin of new properties through the interaction of components of a system. So, if previously isolated components come together and interact, a new system with new properties and laws is formed. Protons, neutrons and electrons together in an atom are more than just the sum of their isolated properties; they abruptly form *elements*. And these are surprisingly different: If, instead of two protons (helium, a gas), we find three together in an atomic nucleus, we have lithium, a solid metal, in our hands. If we add a 17th proton to a nucleus with 16 protons (sulfur, a solid), we suddenly have a gas (chlorine) again. For the sake of simplicity, I have left out the fact that, in addition to the protons, a number of neutrons and as many electrons as protons populate the atoms of the respective elements.

An equally beautiful example are organisms that we call *lichens*. These are symbiotic communities, living beings that are formed from the symbiosis between certain fungi and algae. But this is only a very rough simplification, in reality, lichens are even more complex beings that join together with other additional fungi and bacteria as active participants in the dynamic system. The lichen researcher Goward says: "The only people who can't see a lichen are lichenologists. They look exclusively at the parts, as their training as scientists dictates. The only difficulty is that *you can't see the lichen itself* when you're dealing with its parts."[8] And what you don't see then is: Lichens are incredibly tough organisms, they can withstand years of drought and the strongest Antarctic frost, they can't be killed in climate chambers that simulate Mars nor in outer space. But their various components alone can't survive there.

We cannot derive the laws of evolution from the study of the cells of all possible organisms, not even if we knew *everything* about the properties of elementary particles and could use them for calculations in mega-fast supercomputers. The laws of quantum mechanics

---

[6] W. Ebeling, R. Feistel, *Chaos and Cosmos—Principles of Evolution* (Spektrum Akademischer Verlag 1994), p. 24.

[7] https://de.wikipedia.org/wiki/Emergenz

[8] Quoted from Merlin Sheldrake, "Verwobenes Leben", Ullstein 2020, p. 128, translated by the author.

neither directly nor indirectly create the laws of evolution, which arise and act on a much higher level of aggregation of matter. When life emerged with the inheritance of characteristics through DNA and RNA, when the living organism and the inanimate environment together had formed ecosystems, only then the laws of evolution emerged for the first time. There is no evolution on the quantum level, nor on the level of atoms or molecules.

Now, we take a step further: We also cannot therefore derive the laws that apply to the universe as a whole from the laws of quantum mechanics. With these statements, we have now entered deeply into controversial areas of science and epistemology, which are associated with the concepts of *reductionism* or *holism*.[9] Henning Genz once formulated a position on this that I do not share. He wrote: "Reductionism describes an understanding of how the world actually works; *emergence* stands for our lack of insight into this functioning."[10] This reflects the view of many physicists, who say: If we observe new properties and laws on a higher level of complexity, they would only appear to us as emergent, i.e., as emerging only through the interaction of the components of a system, because we would not have fully understood the isolated components. We will deal with this topic to a greater extent later. My opinion, however, is that emergent properties and laws of complex, higher-organized systems cannot be explained from the properties of their components, but arise through interaction. I will give further convincing examples.

How does it come about that things come and go, living beings grow and die; that cultures arise, mature, become strong and decay; and that we ourselves are conceived and born, become bigger, older and more mature, only to die at some point?

Prigogine has generally explained this to us in a fundamental way with his non-equilibrium thermodynamics.[11] We will look at this again in detail here: While, on the level of the entire universe, entropy can only increase, it can decrease in open subsystems. The entropy generated in such systems is exported out of them—ultimately, into the universe, most likely into black holes. This hypothesis holds weight because, according to a study recently published by the group around Maximiliano Isi, the surface of a black hole does not become smaller.[12] In proving this, the research group confirmed Stephen Hawking's so-called *black-hole area theorem*, which, in turn, leads to an interesting conclusion for our question here: Since the entropy of a black hole is proportional to its surface, the entropy can only increase. And since, as far as has been found so far, black holes constantly absorb matter from their environment, their entropy increases—they thus represent the countless

---

[9] Cf. https://www.spektrum.de/lexikon/philosophie/reduktionismus/1751 or https://de.wikipedia. org/wiki/Reduktionismus or as an opposing position https://de.wikipedia.org/wiki/Holismus; Ebeling/Feistel describe the fact of emergent properties and laws as *irreducibility*, a. a. O. p. 64 and 65.

[10] H. Genz, loc. cit., p. 284., translated from German by the author.

[11] https://www.nobelprize.org/nobel_prizes/chemistry/laureates/1977/prigogine-lecture.pdf

[12] https://www.livescience.com/hawking-theory-confirmed.html, Original publication: Maximiliano Isi et al., *Testing the Black-Hole Area Law with GW150914*, Phys. Rev. Letters, **127**, 011103–1 July 2021.

entropy sinks in the universe. According to a recently published study, it can be assumed that the universe contains 4×10^{19} (or, in other words: 40 trillion) black holes, i.e., ten times as many as galaxies.[13] And more and more will be formed. Which is not the whole story: Primordial black holes might still be around in even much higher numbers and could explain the strange behaviour of galaxies which led to the hypothesis of "dark matter".[14] The readers may want to remember this a little later below when I am referring to a theory by Prigogine and coauthors which is an alternative to the "big bang" hypothesis.

The entropy decrease in the various non-equilibrium systems on Earth and throughout the universe is equivalent to the formation of structures there. In Prigogine's words, these are *dissipative structures* that arise in all such open systems, through which an amount of energy is *pumped* that is greater than critical, as the non-equilibrium thermodynamicists say. This leads to the export of entropy, ultimately, into black holes, and thus to self-organization in the respective non-equilibrium system. In Haken's *synergetics* words, it is the interactions of the system components that cause self-organization. Both are appropriate wordings. Order arises spontaneously from chaos. This is the explanation that the writer Henry Miller was missing when he complained in his novel *The Cancer's Turning Point*: "Chaos is the word we have invented for an order we do not understand."[15] Non-equilibrium thermodynamics leads to this understanding that Miller did not yet have. But it is a remarkable discovery for a non-scientist to see that there is indeed a kind of order in chaos or that it can arise from it.

Do Parlor Games Adequately Represent the Game of Chance and Coincidence in Nature?

In Chap. 2, we have already dealt briefly with the book by M. Eigen and R. Winkler *Laws of the Game: How the principles of Nature' Govern Chance*. When studying this book, one could get the impression that nature is ultimately also pursuing something like dice or strategic parlor games on the way to equilibrium. But this is not the case and would probably be a misunderstanding of the book's message. In fact, it refers, in rather too few and very hidden hints, to the fact that, in complex processes, *dissipative*

---

[13] https://www.eurekalert.org/news-releases/940381, Original publication here: https://iopscience.iop.org/article/10.3847/1538-4357/ac34fb.

[14] Did black holes form immediately after the Big Bang? and are they still around? cf. https://www.esa.int/Science_Exploration/Space_Science/Did_black_holes_form_immediately_after_the_Big_Bang; original publication: N. Capelluti, G. Hasinger, P. Natarajan, "Exploring the High-redshift PBH-ΛCDM Universe: Early Black Hole Seeding, the First Stars and Cosmic Radiation Backgrounds" in The Astrophysical Journal, Vol. 926, No. 2, 205, published 2022, Feb. 25, cf. https://iopscience.iop.org/article/10.3847/1538-4357/ac332d (open access)

[15] Quoted from "DIE ZEIT" No. 7, 11.2.2021, p. 33, re-translated by the author

structures are formed.[16] For example, the ecological dynamics among grass (which is eaten by rabbits and increases their population), foxes (which eat the rabbits) and hunters (who hunt the foxes) is compared with the game *Struggle*. Its rules allow a "game course of action that realistically simulates the temporal change of the different species or their quantities in an ecologically closed space".[17] But one must be careful not to confuse the course of such games and their rules with the actual dynamic rules in real ecosystems. Because these are not at all closed, but are rather open systems that exchange energy/entropy and substances with the environment. Nor are there only four *players* in real ecosystems, but even more, namely mice, birds of prey, insects, rain and many additional actors, not to mention the sun supplying the energy from outside.

Part I of Eigen and Winkler's book, encompassing chapters 1 to 5, is entitled *The Taming of Chance*. Chap. 1 is introduced with [retranslated by myself]: "The game is a natural phenomenon that has guided the course of the world from the beginning: the shaping of matter, its organization into living structures, and the social behavior of human beings." I cannot agree with this formulation—with all due respect to the authors. What we know as *games*, such as dice or strategic board games, follow selected, fixed rules. Rules that we would not find in nature in any way, shape or form. Ultimately, the book is based on *statistical equilibrium behavior*, as can be seen in the *dog/flea* model described there and the numerous zero-sum games cited in the book. But the essential chance and coincidence events that interest us are not statistically analyzable, and the systems in which they occur are all non-equilibrium systems. And chance or coincidence cannot be tamed, unless the non-equilibrium systems slackens and slides into equilibrium.

Another questionable approach can be found in Sect. 4.3, *The law of large numbers*. Questionable because the book is supposed to be about the control of chance/coincidence by natural laws. Of course, the individual processes in large systems are averaged out statistically (the average size of men on the one hand and women on the other; the probability of airplane crashes or disasters in nuclear power plants; the frequency of mutations). But far outside of 99.9999…% of the cases that make up the average, there are essential coincidental events that are significant or even catastrophic as such as well as for the participants or those affected: Whether it is the achondroplastic man (suffering from dwarfism) whose parents are of average size and who accidentally has a corresponding genetic defect, and therefore has to overcome many problems in his life. Or the passenger who arrives too late at the airport, misses her flight and is unbelievably lucky because the airplane crashes by chance due to a collision with a flock of birds,

---

[16] e.g. on p. 114 ff. in M. Eigen, R. Winkler, loc. cit.; it would certainly also be a misunderstanding because Manfred Eigen had developed a model of self-organization, in this case for the origin of life; the model is known under the term "hypercycle", cf M. Eigen, P. Schuster, "A principle of natural self-organisation", Part A: Emergence of the hypercycle, in: Naturwissenschaften, 64, pages541–565 (1977) (English), available here: https://link.springer.com/article/10.1007/BF00450633.

[17] M. Eigen, R. Winkler, loc. cit. p. 110 ff.

while a passenger on the waiting list moves up, was happy that he could still get home on time, but now didn't get home at all. Or whether it is the residents of the area around the nuclear power plant in Fukushima—the people affected by the essential coincidental accidents in each case are not interested in the statistical average of the large number. For the man with dwarfism, it is irrelevant that men in the USA are, on average, 1.754 meters tall. For the man on the waiting list who moved up, it is irrelevant that flying is much safer than driving by car. And when it comes to the fact that it is actually extremely rare, on average, for nuclear power plants to experience catastrophes, especially, by definition, not an accident that is bigger than a DBA, that is, bigger than a *maximum credible accident,* as it nevertheless happened, the affected people are not interested at all.

All such serious or significant, *essential* coincidences—like mutations or, even more so, the origin of a new species—disappear in the "normal Gaussian distribution" curves because of the law of large numbers, that are, indeed, very far outside on the right or left, but they nevertheless represent exactly the kind of chances, coincidences and accidents whose causes we want to understand. The types of events that move the world or our lives can not be taken into account in the frequency distributions. Decisive changes in direction or quality are caused by events that do not occur in probability calculations at all, because they are so extremely rare. However, in Eigen/Winkler's book, chance is only seen as a normal fluctuation of the components of systems, which would then lead to a deterministic behavior according to the law of large numbers, corresponding to a deterministic law and being statistically investigatable. The truth is, if we really want to understand essential chance and coincidence, we must not confuse the statistical average of many events with the causes of the individual events. But the latter is what interests us here in this book.

Although, of course, most systems and phenomena can also be investigated statistically or mathematically with chaos theory, this *does not explain* to us where the chance in the individual case comes from, nor what the *causes* of chance and coincidence are in general! *Climate,* for example, is a representation of the means of individual *coincidental* weather events over a long period of time. A mathematical representation, even one based on chaos theory, *describes what has happened,* but *does not explain* the causes. In addition, a statistical or chaos-theoretical description only applies to those systems that are relatively stable over the observation period. This does not mean *in equilibrium,* but only *relatively* stable. The switch of a warm climate period into an ice age, the appearance of a new type of *Homo sapiens* two or three hundred thousand years ago, the disappearance of a long-reproducing family of species such as the large predators and the plant-eating dinosaurs, the ultimate development of the ancestors of the pterosaurus (flying dinosaurs) into the birds we know today, an invention such as the *World Wide Web,* a composition like Beethoven's *Ninth*—all this and many more are *events,* non-linearities that are not at all to be found in the noise of the statistical averaging of previous events, and not even far out at the ends of the normal distribution curves.

Such, as well as other, sudden events, i.e., breaks in relatively continuous development processes, are also referred to as *symmetry breaking.* Classic examples of this are the almost complete domination of matter over antimatter and the accidental formation of optically active organic substances during and shortly after the origin of life. Eigen and Winkler comment on this as follows: "Symmetry breaking indicates gaps in our understanding of fundamental relationships."[18] Snowflakes are mentioned and shown as examples of the high symmetry of nature,[19] however, these natural works of art are only approximately symmetrical and therefore, in reality unsymmetrical. They are excellent examples of non-equilibrium processes, as we discussed in Chap. 4; symmetry breaks do not point to our ignorance of the fundamental processes of nature, but are essential elements of all processes in our non-equilibrium world.

Weather and Climate are Chaotic

We have seen that structures such as snowflakes or oak leaves, two adjacent trees of the same kind or the clouds of a certain weather situation have the same *pattern*, but are never *identical*. They are also not symmetrical, although it may look like that at a first glance. In detail, they are not. It is characteristic of non-equilibrium systems that we can never exactly predict their course, their development, their future—and sometimes not even roughly correctly. When and where snowfall or rainfall will happen can only be predicted correctly for very short periods of time—"tomorrow and maybe the day after tomorrow"—with a probability of about 60%, beyond which the rate of correct predictions decreases dramatically. Weather is a very typical example of complex non-equilibrium processes in which simply too many non-linear processes interact with each other to make a prediction with 100% or even approximately 100% probability possible. This is not the case because the weather *prophets*, as we sometimes say ironically, are incompetent, or because their computers are not powerful enough or the programs are not well structured—no, it is the fundamental property of non-equilibrium processes and systems to be incalculable. The sentence "The flapping of a butterfly's wings in Brazil can cause a tornado in Texas"[20] is widely known, but mostly not really understood. It serves rather

---

[18] M. Eigen, R. Winkler, loc. cit. p. 133 ff.

[19] M. Eigen, R. Winkler, loc. cit. p. 125.

[20] Edward N. Lorenz, "Predictability: Does the flap of a butterfly's wings in Brazil set off a tornado in Texas?" Title of a presentation in 1972 at the annual meeting of the American Association for the Advancement of Science; Science 320, 2008, p. 431. Here a short portion should be cited: "Errors in the finer structure, having attained appreciable size, tend to induce errors in the coarser structure. This result, which is less firmly established than the previous ones, implies that after a day or so there will be appreciable errors in the coarser structure, which will thereafter grow just as if they had been present initially." He had observed that apparently irrelevant differences in starting parameters resulted in significant differences in the final result of his calculations; cf. https://math-sciencehistory.com/wp-content/uploads/2020/03/132_kap6_lorenz_artikel_the_butterfly_effect.pdf (downloading Lorenz' article); and here is an analysis of the origin of this metaphor: https://www.researchgate.net/publication/26293156_The_Butterfly_Effect_of_the_Butterfly_Effect

as an *intelligent* contribution to the conversation at boring parties in order to show how widely educated one is, or to apparently humorously conceal how bad one's basic scientific knowledge is. What is mostly not understood is the meaning of *can* in this famous sentence: A tiny action can perhaps, but does not necessarily, have to have a big effect.

Short-term weather forecasts are usually relatively reliable, but unstable weather situations can cause the forecasts to be drastically wrong.[21] Only if we could measure much more accurately with much denser arranged measuring stations—one would theoretically have to measure within every cubic meter of air of our atmosphere and within every cubic meter of water of the oceans and lakes—could we possibly predict the weather almost exactly: It would then be eternal night, it would be cold, we would all die, the surface of the earth would approach a state of equilibrium—with maximum disorder, without life (at least, higher life), without dynamics, without further changes; without surprises, without any coincidental events or accidents.

What applies to the weather applies to a much greater extent to the climate.[22] The climate models used so far therefore understandably have a number of weaknesses, although there are several approaches that try to simulate the various non-linear climate phenomena of recent and older history. But they fail, either in the past or in the present. For example, it is not yet possible to simulate the medieval climate optimum or the earlier one during the Roman period, just as the simulations are far from even closely reproducing the Little Ice Age from the 15th to the 19th century in the simulations. Newer models have to prove themselves first, but, so far, they have not made any progress in this respect. However, it is clear that the sun holds a significant influence, which has so far been largely overlooked.[23] It is becoming increasingly clear that there have been different phases of a very dynamic climate since the Ice Age, as evidenced, for example, by the equally dynamic history of glaciers in the Alps, which have retreated at least 12 times since the end of the Ice Age; additionally, during the Roman climate optimum, the Alpine glaciers were significantly smaller than they are now.[24] Newer studies are trying to understand the approximately fifteen-year warming pause from 1998 to 2013 (often

---

[21] In Wikipedia https://en.wikipedia.org/wiki/Weather_forecasting it says: "The inaccuracy of forecasting is due to the chaotic nature of the atmosphere ... forecasts become less accurate as the difference between current time and the time for which the forecast is being made (the range of the forecast) increases ... when in a fluctuating pattern, it [the forecast] becomes inaccurate ..."

[22] https://www.spektrum.de/news/simulationen-wie-ein-klimamodell-entsteht/1781331

[23] https://www.epa.gov/sites/default/files/2014-11/documents/climate_change_and_its_causes_a_discussion_about_some_key_issues.pdf

[24] https://www.researchgate.net/publication/201169725_Multicentury_glacier_fluctuations_in_the_Swiss_Alps_during_the_Holocene

referred to as the *hiatus*). A far better understanding would be possible if natural climate cycles (ultimately driven by the sun) and volcanic events were included.[25]

The non-linearity of the Earth's climate becomes even more pronounced when we extend the period under consideration. The accompanying figure shows 500 million years[26] (homo sapiens has only existed for two or three hundred thousand years). Here, we can see that the climate fluctuations of the last two thousand years were minimal in comparison to those of the last half million or even the further five hundred million years shown. At the same time, we can see that the temperature has been rising for the last 150 years, in parallel with the massive increase in industrial and domestic activity by humans.

---

[25] https://scitechdaily.com/climate-models-cant-reproduce-the-early-2000s-global-warming-slow-down-scientists-explain-why/ and https://journals.ametsoc.org/configurable/content/journals$002f clim$002f29$002f3$002fjcli-d-15-0063.1.xml?t:ac=journals%24002fclim%24002f29%24002f3% 24002fjcli-d-15-0063.1.xml&tab_body=fulltext-display

[26] https://muchadoaboutclimate.wordpress.com/2013/08/03/4-5-billion-years-of-the-earths-temper-ature/, original graphic from https://commons.wikimedia.org/wiki/File:All_palaeotemps.png.

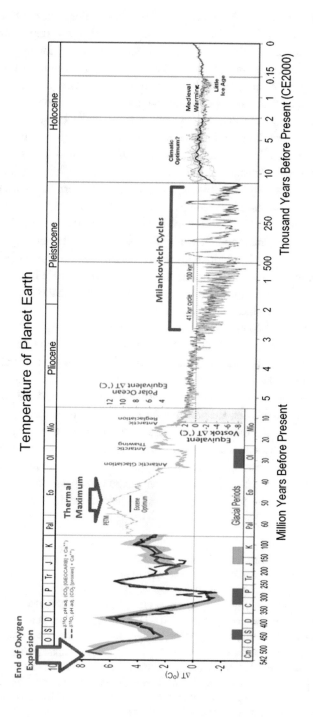

I would like to make something clear here: The climate does not develop linearly, any more than the weather does! Only. the climate is even *more* unpredictable, because there are simply too many different influences that non-linearly interact with each other and are already non-linear in themselves. Let's take the formation or disappearance of clouds, for example: When, at what height, and in what density clouds form or dissolve is extremely difficult and unreliable to calculate. Even less is known about the effect of cloud cover on the average decade- or century-temperature caused by solar radiation, apart from the fact that the extent of the influence of water vapor in the atmosphere was only recently discovered, water vapor, which is not condensed into droplets in clouds, but invisible as *relative humidity*.[27] In addition, there are volcanic eruptions that we cannot even predict a few days in advance with our most modern scientific methods (and, as to the backward simulation, as well as for future climate models, we know practically nothing about them). There are simply too many different non-linear influencing factors, not least the fluctuating radiation intensity of the sun—this does not only mean the infrared radiation, but also the UV radiation, which is often, but wrongly, neglected. Even the orbit of the sun in our galaxy has enormous effects on the Earth's climate, and all these factors interact with each other again non-linearly, and thus make climate development fundamentally unpredictable.

A recent article in *Science* shows how much the climate is also influenced by factors that play no role at all in today's climate models–not even in those that try to include a significant influence of the sun on climate changes:[28] The sudden reversal of the Earth's magnetic field about 42,000 years ago, after which the magnetic field's strength at the end of a longer decline was close to zero, caused global climate change. We should probably say *climate catastrophes*. Responsible for this was the enormous increase in cosmic radiation that reached the Earth unhindered, also further amplified because a solar minimum coincided by accident. This was followed by the loss of the ozone layer, droughts in Australia, with the accompanying loss of local megafauna, dramatic changes in the monsoons, cooling in the Northern Hemisphere. The disappearance of the Neanderthals and the emergence of the first cave paintings are also associated with this event.[29] It is

---

[27] Water vapor is the strongest factor (stronger than CO_2) in the greenhouse effect on Earth, see e.g. https://www.nasa.gov/topics/earth/features/vapor_warming.html.

[28] A. Cooper, C. Turney, J. Palmer, A. Hogg, M. McGlone, J. Wilms and many others, "A global environmental crisis 42,000 years ago", Science (19. 2. 2021) Vol. 371 (No. 6531) 811–818, see https://science.sciencemag.org/content/371/6531/811, summary here: https://www.sciencealert.com/earth-s-magnetic-field-flipped-42-000-years-ago-with-dramatic-consequences.

[29] There are also hypotheses that the eruption of a supervolcano in southern Italy could have killed hundreds of thousands of Neanderthals, so that this eruption could at least have accelerated the disappearance of the Neanderthals. See J. Krause et al, A genome sequence from a modern human skull over 45,000 years old from Zlatý kůň in Czechia., June 2021, Nature Ecology & Evolution 5(6):1–6. The volcanic eruption mentioned here took place approximately at the same time as the previously mentioned reversal of the Earth's magnetic field.

assumed that a reversal is imminent because the magnetic field has been continuously weakening for about 200 years. The behavior of the Earth's magnetic field is another extremely dynamic process, part of many non-linear processes of the non-equilibrium system Earth, which, in turn, has non-linear effects on the climate.

"It is difficult to make predictions, especially about the future." This quote is often attributed to Kurt Tucholsky or Karl Valentin, perhaps because it reflects their way of talking. But it may have been the physicist Niels Bohr who coined the phrase in a speech at a quantum physics conference, or, for that matter someone else, namely the unkown creator of a Danish proverb.[30] Wherever this wonderful sentence comes from, it is true, and an article in the German weekly newspaper *DIE ZEIT* has shown this with many examples: "Economists were as surprised by the oil crisis as politicians were by the 1968 protests and the environmental movement. Instead, in the 1960s, it was expected that mines would soon be operated on the moon. And as to the reunification of Germany that would take place in 1989, a few months earlier, the experts considered it … unlikely."[31] How climate change will actually affect our world is still uncertain. The **IPCC** itself wrote about this in its **2001 report: "In climate research and modelling, we should recognise that we are dealing with a coupled non-linear chaotic system, and therefore that the long-term prediction of future climate states is not possible."**[32]

In my opinion, however, it is indisputably clear: We humans consume too much energy, especially that from fossil sources, we consume too many raw materials, waste valuable drinking water, overuse the soil with industrial agriculture and still throw away a third of the food produced worldwide. We use ever larger areas for cities and infrastructure, destroy far too many ecosystems and annihilate animal and plant species. This will have disastrous non-linear consequences if we do not change course.

---

[30] https://quoteinvestigator.com/2013/10/20/no-predict/#return-note-7474-3.

[31] https://www.zeit.de/2019/38/digitalisierung-fortschritt-prognosen-forschung-innovation/komplettansicht.

[32] https://archive.ipcc.ch/ipccreports/tar/wg1/505.htm, the complete document can be downloaded here: https://www.ipcc.ch/site/assets/uploads/2018/03/TAR-14.pdf. In another report, the IPCC wrote: **"Internal variations caused by the chaotic dynamics of the climate system may be predictable to some extent. Recent experience has shown that the ENSO phenomenon may possess a fair degree of predictability for several months or even a year ahead. The same may be true for other events dominated by the long oceanic time-scales, such as perhaps the NAO. On the other hand, it is not known, for example, whether the rapid climate changes observed during the last glacial period are at all predictable or are unpredictable consequences of small changes resulting in major climatic shifts."** https://archive.ipcc.ch/ipccreports/tar/wg1/045.htm.

Chaos in Our Solar System!

Even processes that run linearly over very long periods of time, such as the orbits of the planets around the sun, can suddenly lead to a significant random event: the impact of an asteroid on Earth. Perhaps its flight path was fairly linear for hundreds of thousands or millions of years, but then it was thrown off course by a close encounter with Jupiter, Neptune or Uranus a long time ago, moving onto another elliptical orbit that ended with an impact on Earth hundreds of thousands of years later. Someone might object that such orbits are at least theoretically quite predictable. We can *foresee* or calculate such events, even if they span very long periods of time and *foresight* is irrelevant for practical purposes. Right, almost completely correct, except for the fact that, because of the *three-body problem*, the orbits of planets and, even more so, the orbits of asteroids and meteorites definitely behave chaotically.[33] This is even more true for the orbits of our planetary system, which is a *N-body problem*.[34] They can no longer be described exactly over long periods of time. We do not know when our solar system will fly apart. Such behavior is called *deterministic chaos*[35], because the laws, here, the laws of motion, are known as such, but the interactions create the chaos. Such things also fall under our definition of a *Deterministic coincidental event*. The orbits of the asteroids and the earth are not predictable, even if it takes a long time to notice that a prediction was wrong … Only then would the chaos in our solar system become slowly visible. This was already shown a long time ago by the mathematician Poincaré, and a simple example is demonstrated very nicely in a video.[36] If even such simple systems are not calculable in principle, it is certainly understandable that more complex processes, in which nonlinearities and all kinds of complex interactions occur, are not calculable. Not because our computers are not big and powerful enough yet, but because these processes are, *in principle,* not calculable.

Back to the different impacts of meteorites and asteroids on Earth. The consequences of such impacts are unpredictable as well, such events brought about the destruction of what existed and the creation of something new at the same time.[37] Did asteroids bring water to Earth? And was the asteroid that struck the Atlantic Ocean off Mexico more than 60 million years ago responsible for the extinction of the dinosaurs, or only of those dinosaur species that did not evolve into today's birds in the course of evolution, which gave mammals and eventually humans the space for evolutionary development? Surprisingly, a new research paper puts forward the very well-founded hypothesis that

---

[33] https://en.wikipedia.org/wiki/Three-body_problem

[34] https://en.wikipedia.org/wiki/N-body_problem

[35] https://www.britannica.com/science/chaos-theory and deterministic-chaos.

[36] http://www.clausewitz.com/mobile/chaosdemos.htm,https://vimeo.com/11993047

[37] Manfred Gottwald, Thomas Kenkmann, Wolf Uwe Reimold, "Terrestrial Impact Structures: The TandDEM-X Atlas", Verlag Dr. Friedrich Pfeil 2020 (2 volumes, English).

this impact was the trigger for the development of the Amazon rainforest as we know it today, as it still exists in parts.[38]

Chaotic Typhoon

Normally unpredictable events happen when at least one non-linear process reaches the point of its development curve, at a bifurcation, where it takes a completely different course. This is much more often the case when two or more non-linear processes interact with each other—things become completely messy. Towards the end of my 13 years in China, in September 2017, I continued working on first drafts of this book in my Shen-Zhen apartment.

I lived on the coast of this southern Chinese city with a free view onto *ShenZhen Bay;* across the bay, I could see the uninhabited coast of an island belongig to Hong-Kong. In the days when I was working intensively on a predecessor to this chapter, a typhoon threatened to come very close to our coast, directly from the sea opposite my balcony. Since I had already experienced several typhoons there, and had once even had the eye of a typhoon pass directly over my residential area—a literally breathtaking and ear-deafening experience! fortunately without any dramatic damage—I always followed the typhoon forecasts with great interest. Over a period of ten days in September, two typhoons had already occurred, and a third one was predicted to be on its way to HongKong and ShenZhen. While the first two landed at locations on the coast of the province of GuangDong, to which ShenZhen belongs, each within fifty kilometers of the originally predicted locations, the third typhoon behaved in a very unfriendly manner. Its arrival was announced for Sunday, September 3, 2017.

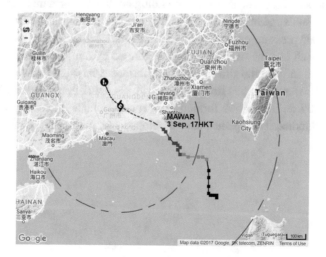

[38] https://www.si.edu/newsdesk/releases/how-chicxulub-impactor-gave-rise-modern-rainforests

On Friday, it was said that *Mawar* would land on Sunday at noon about fifty kilometers west of ShenZhen. But then, it suddenly made a 90° turn to the north (the black part of the route in the picture on the previous page).[39] Now, it was said that it would land one hundred kilometers east of ShenZhen. However, the typhoon did not adhere to this suggestion either, making another 90° turn, this time to the west (the green section of the route). Now the landing was one again predicted to be west of ShenZhen. In addition, the speed of the typhoon was much lower than expected. Finally, the typhoon made another change of direction of about 45° to the northwest and reached the coast about one hundred kilometers east of ShenZhen, but twelve hours later than expected, namely, only on Sunday night. Meanwhile, the power of the typhoon had decreased significantly. It was downgraded to the category of a *strong tropical storm*—that was the complete opposite of the forecast, which, because of the particularly low speed over the warm water, had announced an upgrade to *dangerous typhoon*.

Prigogine showed in his work, which he published essentially in the 1950s and 1960s, that it is the non-equilibrium properties that make our world not (exactly) calculable. We can give probabilities or scenarios for further development, although sometimes not even that. But whether our estimates were correct, we only know when the time comes. It is simply the case that open systems—like the earth and all local sub-systems down to single cells and snowflakes—, which experience an import of energy and matter above the critical point, always behave chaotically. They are not predictable in detail and often not even qualitatively, being full of dynamics, interesting structures and coincidences. Just the way life plays out.

In the case of the earth, the sun provides light and heat, and, together with the residual heat in the earth's interior, which arises from the enormous flow of matter that formed the earth, these are the forces that drive the dynamics in and on earth: The continents move across the earth's surface, volcanoes and colliding continents form mountains and mountain ranges, the sun's radiation causes water to evaporate from the oceans, clouds form. Differences in air pressure cause wind and storm, rain falls here or there from time to time. The rain itself and the rivers, which were formed by rain, carry away the mountains that were formed in primeval times. Organic substances have given rise to life, grasses, flowers, trees, mosses, fungi and lichens grow with the help of photosynthesis, which uses sunlight, whereby fertile earth is formed from rock. Animals, including humans, which live directly and indirectly on plants, have developed and, in turn, shape the earth.

---

[39] Screenshot of the then-forecast course in http://www.hko.gov.hk/en/index.html.

Dynamic Networks

A particularly interesting example of complex dynamic structures is described by Merlin Sheldrake.[40] It is about networks in which mushrooms play a key role: Certain species, the mycorrhiza, associate with almost all plant roots everywhere, even in the forest, a fact that led to the catchy moniker *Wood Wide Web*. Such networks cannot be described by listing or mapping their participants, their nodes. You cannot study them by cutting them. In that case, you still wouldn't know what the network *does,* because a mycorrhiza-root network is not an object, but a living being (rather, several living beings), a dynamic process. They are *complex adaptive systems.* Just as we cannot understand the Internet and its networks by analyzing the end devices such as laptops or smartphones, the servers, the copper and fiber optic cables isolated from each other, and possibly overlooking the mobile connections because they are not visible.

In general, we cannot understand the complexity of higher levels of organization by breaking them down into their individual components and analyzing those in detail. We thus miss the interactions! If aliens were observing and researching us, they might be wondering: "What is the Internet for, what are all these cables and laptops and servers and satellites for?" In addition to mobile networks, they have probably also overlooked the actors who use the end devices to exchange messages and information, or simply play games or watch movies. Even today, we do not know what the fungi are actually *doing* with the plants in their gigantic invisible underground networks. And vice versa, we do not know what the plants are doing with the fungal network. Well, the fungi transport substances from A to B and C in their mycelia, sometimes in the opposite direction; they also seem to be able to transport information—by the way, both much faster than with diffusion speed, so it is a kind of active transport, the mechanism of which is not yet understood. But how do the apparently randomly growing mycelial threads find out that a thread has found a worthwhile target somewhere else, how do the other threads *know* that they can now withdraw? Why don't they stay there and keep looking? We can only find out if we examine the network as a whole and with its activities, with its dynamics, instability and ability to adapt to change. As difficult as it is, and we are obviously still far from understanding this and practically all other networks in nature.

Network formation is one of the many typical examples of self-organization that occur in non-equilibrium systems and processes. We find them everywhere in nature, and thus also in our society. They are an inevitable product of energy input and entropy export, and they arise without us having to take care of them: completely on their own. This is spontaneous self-organization. We can observe it in our human environment. It starts in kindergarten, continues on the street, at school, at work, in sports clubs and parties. Small and sometimes very small sub-networks form within large, coarsely woven networks. We can see it in the animal kingdom, in plant communities, and some of them we call *ecosystems.*

---

[40] M. Sheldrake, Entangled Life (Verwobenes Leben, Ullstein 2020), loc. cit. pp. 246 ff.

Similar, but Not Identical Patterns

In the world of the living, we find patterns that are comparable to those in the non-living part of nature: Just as no snowflake is *identical* to another one, so the DNA of individuals of the same species is not really *identical*. There are always some—be they extremely small and fine—deviations. This is even the case with identical twins, although only in detail, but it does mean that they are not actually 100% identical. This was shown by Julia Freund et al in a *Science* article headlined "Emergence of Individuality in Genetically Identical Mice": 40 genetically identical mice grew up together in a big joint space with many interesting opportunities for exploration and activities, i. e., all of them, being genetically identical, experienced the same environment; nevertheless, each animal behaved individually differently and ultimately had different body weight and brain volume.[41] Monozygotic twins do not even have identical fingerprints (apart from the fact that their DNA is not 100% identical anyway).[42] What is little known is that, in extremely rare cases, even different genders can occur: If, during an early cell division, one of the male twin embryos loses the Y chromosome due to a n accidental error, the growing embryo only has one X chromosome; thus, it is externally a girl, but suffers from the so-called *Turner syndrome*.[43]

After a woman has had sex with a man—hopefully, full of love, joy and lust—an embryo may develop. In the vast majority of cases, this does not happen, and in the end, it is again a matter of chance whether inception takes place. Whether it will be a boy or a girl cannot be predicted. Only after an ultrasound examination, at the earliest, from the 12th week of pregnancy, can it be determined—if one wants to know—whether it will be a boy or a girl. And even after such an examination, there have been many surprises. But no one can predict how tall he or she will be, how intelligent or practically gifted. Will the child later be a satisfied craftsman or farmer, musician, writer or physics professor, will he or she be successful or frustrated in life, will he or she be homosexual or heterosexual, largely healthy or perhaps chronically ill? No one can predict.

Even if we plant two acorn twins only 20 m apart, where the ground and all environmental conditions—as far as we can judge—are identical, two oaks that develop will not look exactly the same. There will not even be two oak leaves that are exactly the same. The oaks will only have the same *pattern* of their bark, the same *pattern* of their crown shape and their leaves, we can see that they are *oaks* and not aspen or beech. But they will never be *identical*, because the branches and twigs have an infinite number of ways

---

[41] https://www.researchgate.net/publication/236675853_Emergence_of_Individuality_in_Genetically_Identical_Mice

[42] Xi Jinxi, J. D. Glover et al., "Limb development genes underlie variation in human fingerprint patterns", Cell Vol 185, No. 1, P95-112.E18, Jan 06, 2022, see in www.cell.com: https://bit.ly/3xgrCaa, cf. as well in cell.com: http://bit.ly/3Xq1HaQ

[43] https://www.reference.com/science-technology/can-boy-girl-identical-twins-f20c37b6ec0408c1

to branch off. The technical term for *branching off* is *bifurcate.* That is why Prigogine used the term *bifurcation,* introduced earlier by Poincaré, to describe his observation that, with the constant increase of a critical parameter—for example, temperature or amount of energy—systems often reach points at which they have two or more options for further development.[44] It describes the phenomenon when changes in the qualitative state occur in nonlinear systems. Such branchings—the forks in the roads of our lives—allow chance and coincidence to strike constantly. And so, at one tree, the branch branches off here, at the other tree, the next branch branches off there, becoming another branch that allows other branches to branch off in this or that way, until, finally, a unique tree is created—so it is for us with our own road forks: They branch off until, finally, every life becomes unique, repesenting experiences that, in its many details, are shared with no one else in the world, one that is not repeatable and not predictable.

Just as it is unpredictable how a river will shape the banks between which it flows, and how the ground and the banks will shape the river—because this, too, is something reciprocal, based on intense interactions, with an infinite number of possibilities for coincidental effects—two rivers in the same environment may not even be remotely alike. We will hardly ever have the opportunity to see two real rivers in really comparable environments, but, on a small scale, it is possible: The two photos below show two tiny *rivers* on a beach in northern Brittany (France), only two meters apart, and not even these look the same.[45]

 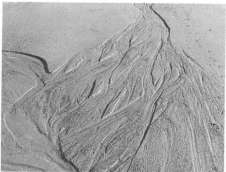

How different each river system is from all others is shown by the fascinating river maps prepared by "Robert", who started *grasshoppersgeography* and, among other things, shows the rivers of the world with their tributaries.[46]

The NASA satellite photo on the cover of this book (the mouth of the Lena river in North Siberia) shows us a very typical non-equilibrium system, a typical dissipative structure. The delta will look different again a little later. The patterns on the wings of

---

[44] https://en.wikipedia.org/wiki/Bifurcation_theory

[45] Photos by the author.

[46] https://www.grasshoppergeography.com/River-Maps/

a butterfly species, the stripes of zebras, the fur patterns of giraffes, tigers and leopards are also the result of dissipative structures in the non-equilibrium system of the diffusion of pigment-forming and the formation of molecules that slow down diffusion or prevent pigment formation. The British mathematician Alan Turing investigated this phenomenon mathematically and was able to simulate such patterns with partial differential equations. After that, these patterns have also come to be called *Turing patterns*.[47] Prigogine designed the chemically (non-equilibrium) thermodynamic basis with a model of reaction equations for the already mentioned Belousov-Zhabotinski reaction, which leads to similar patterns; his model is called *Brusselsator*.[48] While the Turing simulation is only a *description*, the *Brusselsator* model provides an *explanation* for such pattern formation on a thermodynamic basis.

Nothing is as Constant as Change

The description above is not only the case for two people or two oaks, but also for societies, cultures, their history. It is generally true for the entirety of animate and inanimate nature, for continents, mountains and valleys, for the sun and other suns, for planetary systems in the Milky Way, for galaxies and galaxy clusters: All of these are respectively different from each other, all are dynamically changing all the time, their behavior is not predictable in detail, and most of the time, a future trend cannot be derived from the past. Our entire universe is in constant change, both in small and in large scale. Complex structures have arisen and are still arising, *dissipative structures*.

For many people, as can be read uncontested on the German Wikipedia page, "chance" is an event or a coincidence of events for which no causal explanation can be found.[49] In Chap. 2, I have already pointed out that this is not only the view of philosophically or scientifically naive Wikipedia authors: The idea of *Indeterminism* is quite popular. The *Causal* intentions of acting humans or scientific *deterministic processes* could be considered, and if these were not present, what happened, would be a matter of chance. I have already contradicted this idea with the definition used for this book, because, of course, all events, even coincidental ones, have a cause. We often just don't know it. The decisive trigger in the web of non-linearly interacting processes could be the famous butterfly; how will you determine afterwards which butterfly it was? For many Chinese, coincidental events are *fate*. We humans often try to ascribe meaning to

---

[47] https://royalsociety.org/blog/2021/11/turing-theory-pattern-formation/, see also https://www.nature.com/articles/482464a and here Turing's original paper: https://royalsocietypublishing.org/doi/10.1098/rstb.1952.0012.

[48] https://en.wikipedia.org/wiki/Brusselator.

[49] https://de.wikipedia.org/wiki/Zufall.

the course of events, even if it is to suspect supernatural forces or beings at work, which leads to an esoteric world view or superstition.

However, our ever-deeper understanding of the universe could actually show us that we need neither elementary particles with their indeterminacy, nor fate or supernatural explanations for the emergence of the phenomenon of *chance and coincidence*. Even as we only have diffuse ideas about the beginning of the universe, and these ideas are likely to change in the next decades and centuries, it is clear to careful observers: The highly complex and diverse structures in the universe are not consistent with the assumption of a homogeneous beginning—let's call it *The Big Bang* as usual—and the *cosmological principle* associated with it. This principle states that the universe is homogeneous on large scales and looks *equal* in all directions and from all locations. It is a key pillar of the *Standard Model* of cosmology.[50]

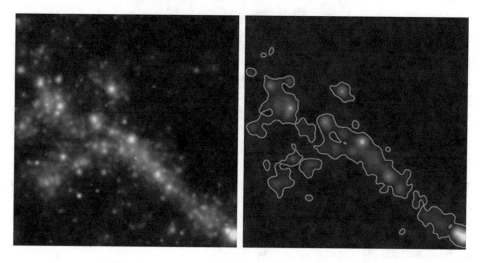

Recently, a worldwide research group collaboration discovered that hundreds of thousands of dwarf galaxies are practically lined up one after the other, as can be seen in the images above and on the next page.[51] The large-scale structure is clearly visible in it: It shows the light emission of hydrogen gas with a diameter of 15 million light-years, with the bright points in the filaments each showing a galaxy in which the first stars are forming.[52]

---

[50] https://www.intechopen.com/chapters/66783.

[51] *R.* Bacon and many others, "The MUSE Extremely Deep Field: The cosmic web in emission at high redshift", Astronomy and Astrophysics 647 (2021) A 107 (26 pages), accessible here: https://www.aanda.org/articles/aa/full_html/2021/03/aa39887-20/aa39887-20.html

[52] Fig. 4 in the press release of the CNRS: https://www.cnrs.fr/en/first-images-cosmic-web-reveal-myriad-unsuspected-dwarfgalaxies; Reproduced with kind permission of the CNRS, researchers Jeremy Blaizot, Roland Bacon and Thibault Garel

Also currently in the news is the discovery of incredibly large and far-reaching gas filaments in the universe, objects that had been detected earlier but on a much smaller scale.[53] These much larger examples were found by a large international team of researchers led by the *Argelander Institute for Astronomy of the University of Bonn*: the total length of the network is 50 million light-years, and that is only the mapped part.[54] These gas filaments are connecting elements of the "sponge structure" of space created during the expansion of the universe. Huge spaces emerged that contain virtually no galaxies, but these empty spaces are, so to speak, enveloped by so-called *walls* of denser arrangements of galaxies, including the Great *Wall*[55] and the *Sloan Great Wall*[56]. The clusters of galaxies and the agglomerations of these clusters in the *Great Walls* are networked together by these filaments. One can practically see their formation as *dissipative structures* in the non-equilibrium system universe. A similar situation exists with billions of dwarf galaxies that are *connected* to each other.[57]

These structures also arise by chance, that is, it is not known or predetermined from the initial conditions of the universe how they will look now. We can not predict how they will look in a billion years. Recently, something even more surprising and hitherto

---

[53] https://science.sciencemag.org/content/366/6461/97

[54] https://astro.uni-bonn.de/~reiprich/A3391_95/A3391-95_2020_12_04.pdf (Pre-publication from "Astronomy & Astrophysics"), a German summary can be found here: http://www.raumfahrer.net/news/astronomie/25122020172213.shtml.

[55] https://en.wikipedia.org/wiki/CfA2_Great_Wall

[56] https://en.wikipedia.org/wiki/Sloan_Great_Wall

[57] https://scitechdaily.com/first-images-of-the-cosmic-web-reveal-unsuspected-presence-of-billions-of-dwarf-galaxies/

inexplicable has been discovered: a *giant arc*. It is 9.2 billion light-years away from us and spans 3.3 billion light-years.[58] The sky-spanning corresponds to 20 full moon diameters. The discoverers write: "In cosmology, we expect matter to be evenly distributed across the observable space when scientists view the Universe on a large scale so there should be no noticeable irregularities. This is based on the principle that the part of the Universe we can see is viewed as a 'fair sample' of what we expect the rest of the Universe to be like. This is known as the Cosmological Principle.The discovery of the Giant Arc adds to the number of structures on scales larger than those thought to be 'smooth' and therefore pushes the boundary size for the Cosmological Principle." We must also be aware that this newly discovered structure existed more than 9 billion years ago, when the universe was only half as small as it is today according to our current understanding. This shakes the *cosmological principle* to its foundations.

Can The Big Bang Theory Explain the Structures in the Universe?

My understanding of an answer to this question is essentially based on Prigogine's theories, which are all, in principle, falsifiable, that is, verifiable, provable or refutable. Let's start with the *Big Bang*. If you search for information about the Big Bang or if you read relevant articles more or less regularly, you may get the impression that the *Big Bang* has already been proven. But that is not the case. If the cosmological principle cannot be upheld, the Big Bang hypothesis is also no longer appropriate. Even more importantly, the model of the Big Bang is interestingly completely incompatible with the theories that led to its formulation. The Big Bang hypothesis arose from the observation that the universe is expanding. Thus it must have started its existence in an infinitely small point and unimaginably high temperatures billions of years ago, as is generally assumed.

This hypothesis has two weaknesses. One is that the infinitely small and infinitely dense starting point of the universe is, mathematically speaking, a singularity, that is, it cannot be described. And secondly, the laws that are to be carried out *backwards* to the Big Bang are continuous, symmetrical and reversible with respect to time. They describe something continuous. There is no way in the formulae that something can *start at once*. So, there can be no *events* according to these laws. A Big Bang would certainly be an event, and what an event it would be!

---

[58] https://www.uclan.ac.uk/news/discovery-of-a-giant-arc-in-distant-space-adds-to-challenges-to-basic-assumptions-about-the-universe

Edgard Gunzig, Jules Géhéniau and Ilya Prigogine have tried to describe the beginning of the universe differently, and have also relied on considerations by Tyron and Hawking.[59]

Before the beginning of everything, there is only the *Nothing,* but this *Nothing* is not really nothing, because it can fluctuate, it is unstable. The Nothing is also called *quantum vacuum,* and it fluctuates in accordance with Heisenberg's uncertainty relation, the basic equation of quantum mechanics. Elementary particles and anti-particles arise, immediately canceling each other out.

Since the amount of energy in the quantum vacuum—in the *Nothing*—is zero, it can also be understood as the sum of negative gravitational energy and positive energy bound in matter:

$$-E\,(\text{grav}) + E\,(\text{mass}) = 0$$

Neither does this total energy sum (zero) change when the *Nothing* is transformed into *Something* by quantum fluctuation, and this *Something* is mass and gravity. Thus, the total energy of our present universe is also zero.

If, according to the hypothesis of Prigogine and his co-authors, the instability increases, tiny black holes arise, smaller than nanoscopically small. Such black holes decay; the smaller they are, the faster, they decay into the quantum vacuum.

Next, the author group introduces another new aspect into the basic laws of non-equilibrium thermodynamics, namely, the generation of matter from gravitational energy. Since there can always be gravitational energy in the instabilities of the vacuum, matter can arise from it. This creates entropy. While *before,* the quantum vacuum had no entropy, it *after*—because there are now particles, that is, mass—it does have a certain—albeit very low—amount of entropy. Suddenly, there is what Einstein describes as *space-time,* the four-dimensional universe with three spatial dimensions and the dimension of time.

Immediately, we have entropy. The author team shows, with other equations, that mass can arise, but cannot be converted back into *space-time.* This theory now contains no singularity, representing a continuous span, without contradiction, from the empty quantum vacuum to the beginning of time, becoming and decaying. And because the entropy is not *maximal,* but, on the contrary, has some extremely low initial value, the universe does not begin—as is often claimed, see Chap. 7—as a homogeneous, hot, structureless soup, but rather as something that already has structures, something that has complexity. (By the way, the theory of primordial black holes by Capelluti, Hasinger and Natarajan[14] can be seamlessly connected with the one by Prigogine et al.: The pri-

---

[59] E. Gunzig, J. Géhéniau, I. Prigogine: Entropy and Cosmology. In: nature Vol 330, 17 Dec 1987; Prigogine, Géhéniau, Gunzig, Nardone: Thermodynamics and Cosmology. In: General Relativity and Gravitation, Vol 21, No. 8, 1989; the article can be found here: https://www.ncbi.nlm.nih.gov/pmc/articles/PMC282204/ and here https://www.researchgate.net/publication/255936551_Thermodynamics_and_cosmology; see also Ilya Prigogine, Isabelle Stengers, "Das Paradox der Zeit—Zeit, Chaos und Quanten", Piper-Verlag 1993 (the English edition "The end of certainty: time's flow and the laws of nature" was published in 1997 by Simon & Schuster NY)

mordial black holes would be part of the complex structures in the beginning of the universe.) This corresponds, on the one hand, to the understanding in non-equilibrium thermodynamics, according to which low entropy is equivalent to structures—that is, the opposite of a state in which all elements of the system are distributed statistically homogeneously in space—, and, on the other hand, to the basic law of thermodynamics, that the *total* entropy of the universe can only increase and will not decrease.

So, the universe already began with certain structures. Meanwhile, it is a largely accepted fact that the universe's microwave background is not, as previously assumed, homogeneous, but structured. This is a clear indication that structures already existed in the very earliest days of the universe (see pictures below).[60] Prigogine could not have known this at the time, but it does confirm his theory to some extent. In addition, the inhomogeneous microwave background is in line with the large-scale structures of the universe observed today. And it is also in line with the fact that the universe contains a multitude of non-equilibrium systems.

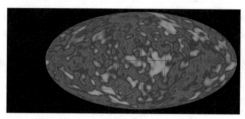 

I will not go into further detail about his cosmological theory here. In summary, it says: The origin of the universe is a succession of instabilities, is irreversible and has always had the character of a non-equilibrium system. Prigogine calls it *phase separation between gravity and matter,* given that the two were indistinguishable in the quantum vacuum. When the phase separation took place, time emerged as well. We will deal with the question of what time is, what the nature of time is, in the next chapter. With the phase separation and the origin of entropy, the universe began with the many events that we call chance or coincidence.

We cannot predict earthquakes or volcanic eruptions, nor, often enough, even not tomorrow's weather. We can not foresee the *Dow Jones* index or the DAX for the next trading day, let alone for Dec. 31st, when we are supposed to predict it in January, not even in the middle of the year or in November. Despite a certain regularity in the cycles

---

[60] https://de.wikipedia.org/wiki/Hintergrundstrahlung; here it is interesting to see that in the accompanying text, it is first said that the microwave background is extremely homogeneous, although the temperature variations can be seen, in a later part of the text it says that the temperature differences are important for cosmology and for theories concerning the structures of the universe. According to Ebeling/Feistel, loc. cit. p. 82, the most important physical causes of structure formation include, among others, the occurrence of gravitational instabilities, inhomogeneities in the initial distribution and entropy release of material matter to the background radiation.

of solar activity—the cycle length is, on average, about eleven years, but with a fluctuation range of seven to fourteen years!—we cannot predict when the next solar minimum and maximum will be reached—and with how many spots.[61] For the current, recently started solar cycle, there are two exactly opposite predictions: One predicts another weak cycle, like the last one, the other a quite strong one.[62] The additional presence of incidental events also in the sun has to do with the interaction of several non-linear processes, just as in other non-equilibrium systems.[63] There are enormous numbers of different currents in the sun, which are probably all turbulent, surface currents and those extending deep towards the center. In addition, there are solar flares, a very dynamic event, so that we should no longer surprised to experience a lot of surprises.

Observe a turbulent flow area in a stream: When does the image you have seen before repeat itself exactly? Actually, never, it is always new, always different in detail, but with a similar pattern. Such currents very often reach *bifurcations,* that is, situations in which there are two or even more possibilities for further development, see also Appendix 12. We see no less bifurcations, that is, branching, in the delta of the Lena River on the cover photo of this book. And take another look at the photo in the foreword: The cirrus cloud shows turbulent patterns similar to smoke in the air—only on a much more gigantic scale! Turbulences everywhere.

The Birth of Chance and Coincidence

Let us finally take a break in our turbulent expedition into a variety of more or less unexplored landscapes of knowledge. We have considered mayonnaise and other dispersions, soccer games, evolution, mushroom networks, river courses and their delta mouths, the weather, the climate, and now also the sun and the galaxies, as well as galaxy clusters and unimaginably large structures in the universe. We can see that practically everything in our life, in our environment and in the universe is a non-equilibrium system or process. We therefore now summarize their most important properties:

- We are talking about open systems, that is, those that can exchange energy and matter with the environment.
- There is an entry into these systems of a supercritical amount of energy into these systems, which drives them far from the equilibrium state.
- This causes a massive export of entropy, which decreases within the systems. Reduction of entropy is equivalent to sudden and spontaneous self-organization into

---

[61] https://en.wikipedia.org/wiki/Solar_cycle.

[62] https://news.ucar.edu/132771/new-sunspot-cycle-could-be-one-strongest-record

[63] https://www.science.org/doi/10.1126/science.aaz7119

dynamic structures that are subject to constant change! In parallel, entropy increases somewhere in the universe, probably in black holes.

- All processes are irreversible; what happened once is done and cannot be undone.
- Nonlinear laws prevail, so that jumps in behavior or in properties occur. If more than one nonlinear influence is effective, these interact and potentiate the number of possibilities for such jumps. Equations, such as those developed by Haken to describe self-organization, are nonlinear.[64]
- At such points, the system therefore shows instability, and the so-called bifurcations occur: The system can develop in different and also surprising directions, depending on the type of specific interactions in the respective system.
- The world of matter is organized in these interlocking systems on successively higher levels: atoms → molecules → solids, liquids, gases; or cells → organisms → ecosystems, societies. And it goes on, with solar systems, galaxies, galaxy clusters and other superstructures that we looked at here by way of example. On each higher level of aggregation of matter, new properties, new laws arise, such as those that cannot be reduced to or directly derived from the properties and laws of the lower level(s).

Chance and coincidence arise at bifurcation points in this complex landscape of non-equilibrium, symmetry breaks, phase transitions and non-linearity of these systems. And that is what our life and the entirety of nature is all about. While structures and processes are formed far away from equilibrium, while entropy is exported in excess, every system always arrives at forks in the road, at which it can take different directions. This leads to coincidental events, especially when two or more causal chains intersect, causal chains that had been independent of each other before, that had developed independently of each other.

The events that happen by chance, coincidentallyly or unpredictably, all have their material cause(s). In principle, this is *deterministic chaos*. All individual processes and the laws underlying them are describable. We may even know at which points, under which circumstances jumps can occur—but, due to the interactions with other processes in the system, even the smallest deviations in the initial or intermediate states—initiated by the often cited and obviously extremely busy *butterfly*—can cause big jumps in properties or behavior, without us being able to predict whether and when and what will happen abruptly—*by chance*.

At this point, I would like to prevent a potential misunderstanding: One could assume that *every event* happens *by chance*. No, not every event happens by chance or is coincidental, by far not every one! But every coincidence is an event. An *event* is, first of all, a non-linearity in a previously linear process of a non-equilibrium process, and most events are predictable or calculable. Let's look at a lunar or solar eclipses: both are events, both are non-linear events in the course of otherwise linear processes. Astronomers can calculate these a long time in advance. The linear orbits of the moon, the earth

---

[64] H. Haken, loc. cit. p. 14.

and the sun make such a calculation possible, and one may expect with certainty that, at the time predicted in advance, the brightness of the moon or the sun will not-linearly decrease and, soon after, increase again. An event is only then a chance or coincidental if two or more completely independent causal chains overlap.

How completely different my life would have been, the history of the company Zipperling Kessler[65] followed by Ormecon,[66] even that of the company EMC, even of the global circuit board market, as well as the lives of hundreds or thousands of factory workers, engineers and scientists, if Wolfgang C. Petersen had not looked at the academic job advertisements in the science section of the *Frankfurter Allgemeine Zeitung* on a certain Wednesday in September 1980 for this single time in his life, where he found my job ad. And he only did that because he was, unusually, having trouble falling asleep and was bored; he had nothing else to read. This was a classic fork in the road, a bifurcation, a coincidence caused by many previously independent, and then interactively non-linear processes. But this coincidence would have had no effect if I had not recognized, at the job interview, that this opened a chance for me to design a task that was fitting for my abilities and interests. Essential coincidences alone do not accomplish anything. Only if you actively search and pay attention, recognize the coincidence, the sudden opportunity, and if you then make a decision in the right direction (even though it only can be determined later whether it was right or not), you can use the coincidence positively. This is then no *lucky coincidence,* but *serendipity,* one of the many types of chance events.

Ebeling/Feistel describe nothing else and also mention the term *coincidence:* "Of principle importance is the connection of entropy and predictability. [...] Each of us experiences this in everyday life. We experience individual chaos in the midst of social stability. [...] Our life's path [...] is characterized by dynamic instability. Moments, moods or coincidences [whereas moments and moods are also coincidences, note by the author] influence our personal fate and set switches for the whole further life. [...] extreme dependence of further development on tiny details is the typical property of chaotic dynamics."[67]

Branches on growing trees happen sometimes here and sometimes there, depending on where the necessary growth hormones accidentally stimulate the growth of the branch cells. Accidentally, because the concentration of the substances responsible for the formation of the branch is not the same everywhere, but, with slight deviations at a random location, the branching then starts there. Here, it is not the flapping of a butterfly wing, but a slight fluctuation in substances that coincidentally initiate the formation of a branch

---

[65] https://www.researchgate.net/publication/262076086_Short_History_of_Zipperling_Kessler_Co_Ormecon's_mother_company

[66] https://www.researchgate.net/publication/260427241_Milestones_highlights_of_the_Organic_Metal_Polyaniline_Science_Technology

[67] W. Ebeling, R. Feistel, loc. cit. p. 62, translated by the author

or twig at the location, with the result that all trees and shrubs of a species—even of one variety and even if they have grown from two offshoots of the same original plant—never look the same, but only similar.

We humans often believe that we can predict the consequences of certain measures. In the summer of 2020, tourism was extremely restricted worldwide, and thus also in Europe, due to the Corona pandemic. Of course, this also affected nature reserves. In many places, it was said that this would be good for the nature reserves and national parks, at least to those that are far enough away from cities. The nature reserve *Duven-stedter Brook* in the north of Hamburg, on the other hand, was at times overcrowded with the result that there was a verifiable disturbance of the crane and deer populations. Without people, or, as in previous years: with fewer people, nature would be better off, it was often written. A study in Sweden surprisingly shows the opposite[68]: The breeding success on a guillemot breeding rock was lower than in previous years. In the end, it became clear that there were seven times more sea eagles in the area than before. The absence of tourist boats allowed the sea eagles to conduct many more search flights than in previous years, which led to the disturbance of the guillemots and a decrease in breeding success. An unpredictable consequence, an accidental result that we understand retrospectively, but would not have expected in this way. The coincidences due to the pandemic caused the birdwatchers to stay away. This was a chance welcomed by the sea eagles to finally give them some undisturbed time to hunt there. But it didn't do the guillemots any good.

Coincidental events are, because their causes are of a material nature, mostly explicable in hindsight: When a strong earthquake suddenly and accidentally occurred off the coast of Japan in early 2011, a tsunami was created. By no means does every earthquake cause a tsunami, but this time, the strongest tsunami in Japanese history developed. It flooded huge areas of the east coast of Japan, killed over twenty thousand people and swept over the safety wall of the nuclear power plant. Together with other coincidences that arose from the past—the planning of the power plant—and the difficult situation, a catastrophe developed. The strength of the earthquake, the location of the epicenter and the V-shaped coast made the devastating effects possible. This nuclear disaster would probably not have come about at all if the epicenter had been only ten kilometers deeper or fifty or one hundred kilometers further north. But it happened as it happened, and the island of Honshu was shifted 2.40 m to the east. Due to the change in mass distribution, the length of the day on Earth shortened by 1.8 ms.[69]

The goal was only scored during the soccer match I mentioned because the leg of the defender accidentally deflected the ball in a different direction.

The earthquake off the coast of Japan in 2011 was a big, in this case, a catastrophically effective coincidence; the stupid goal, however, was a triviality that was quickly forgotten. Both have the character of a non-equilibrium process: plate tectonics or

---

[68] https://www.sciencedirect.com/science/article/pii/S0006320721000021

[69] https://en.wikipedia.org/wiki/2011_T%C5%8Dhoku_earthquake_and_tsunami

soccer; the formation of structures: we observe structure in both processes, i.e., the export of entropy; the non-linearity: the sudden earthquake, triggered after continuously increasing tension between continental plates, or the sudden shot of the striker and the quick reaction of the defender; and from all this, the bifurcation, the coincidence: on the one hand, the earthquake, with tsunami and subsequent catastrophic consequences, the many dead, the radioactive contamination of a large area of land; on the other hand, the shot with the deflection and the subsequent annoying goal, which means the relegation of the now losing team, while a draw would have been enough to stay in the league.

Chance and Coincidence is Everywhere—Just Not on the Quantum Level

These last examples show once again: Chance and coincidence are everywhere. They are reborn in all complex non-equilibrium systems over and over again. They act on large and small scales, from the very small, nanoscale to the extremely large scale of the entire universe. On all these scales, the same driving forces are creating coincidental and accidental events: the export of entropy, the non-linearity of the interacting processes with the possibility of bifurcation, and the fact that, on higher aggregation levels, entirely new laws arise that cannot be reduced to the lower levels or cannot be derived from the properties of the components of the lower aggregation levels.

That is why I deliberately, and for solid scientific reasons, did not include the quantum level: On this level, there is no entropy.[70] Whatever laws may be effective on the level of elementary particles, they cannot cause essential chance or coincidence on the next higher levels. Each aggregation level of matter has its own laws, as I have tried to explain. So, for example, the coincidences in my life have nothing to do with the coincidental events during evolution. Undoubtedly, I am a product of evolution, but the fact that Wolfgang C. Petersen by chance opens an advertisement page of the FAZ on that special Wednesday, a newspaper page that he has never looked at before and never will again, is completely independent of the coincidental character of evolution, or that of the function and the emergence of enzymes in his body cells. These are all completely different levels, of course, ones that still depend on each other, because the higher aggregation levels would not exist without the lower ones! Without evolution, we would not exist, without the cells with their non-linear processes, our bodies would not function,

---

[70] If you look at how entropy is defined (for example, you can find some information about it here: https://www.britannica.com/science/entropy-physics), you will see that all of this is irrelevant for the quantum level: Because on this level, there is no temperature, there are no "state variables" like pressure and volume, no chemical composition. The error in thinking is often made to associate the probabilistic behavior of quantum with chance and coincidence on the macroscopic level. I believe this is not permissible and leads astray.

but the coincidences in my life arise on a completely different level than those in the cells or on the level of evolution.

But how is it that—while uncertainty on the quantum level is supposedly typical for this level—this behavior of quanta (which is more correctly called *indeterminacy* or *indeterminability,* and not *randomness*) *cannot* cause the real, essential chance/coincidence on the higher organizational levels of matter?

I see the answer to this question in so-called *decoherence.* Originally, this was called "quantum decoherence" or, more precisely: "environmental decoherence" or "environment-induced decoherence". This was first introduced in 1970 by the German theoretical physicist H. Dieter Zeh[71], in the publication cited below in the footnote.[72, 73] Of course, Prof. Zeh did not develop this theory in order to explain why the indeterminacy of quanta is not the cause of chance or coincidence in the macroscopic world. Rather, his concern was to answer the question of why, during a measurement, one can never observe the strange properties of quanta that have been set forth, but only something real. Namely: If we have detected the elementary particle at one location, then it was definitely there, and at the same time we have information about where it was not, namely, at another location, where it *could have been* with a certain probability. Later, the scope of this theory was extended to the question of why it is not possible to observe even the very smallest macroscopic systems in their quantum properties, and why, instead, one always measures a macroscopic reality. An introduction into the phenomenon of *decoherence* can be found in the link in the footnote below.[74] Relatively recent overview articles have been written by M. Schlosshauer[75] and G. Bacciagaluppi.[76] "Quantum decoherence plays a central role in the dynamical description of the transition from the quantum to the classical state." Schlosshauer also quotes H. D. Zeh at the end, who wrote in 1997, 27 years after the first publication of his theory: "Decoherence is the normal consequence of interacting quanta. It can hardly be denied–but it also cannot explain anything that could

---

[71] https://en.wikipedia.org/wiki/H._Dieter_Zeh, his publications are (at least partially) accessible here: https://inspirehep.net/authors/982291 and his former home page here: https://www.thp.uni-koeln.de/gravitation/zeh/index.html.

[72] H. D. Zeh, „On the interpretation of measurements in quantum theory", Found. Phys. 1 (1970), 69–76, see https://link.springer.com/article/10.1007/BF00708656

[73] Henning Genz refers to this phenomenon (loc. cit. p. 234 ff., v. a. 242 ff.) as "consolidation", which I also find very accurate.

[74] https://link.springer.com/chapter/10.1007/978-3-540-88169-8_5

[75] M. Schlosshauer, "Quantum decoherence", Physics Reports 831 (2019), 1–57, online zugänglich hier: http://faculty.up.edu/schlosshauer/publications/Schlosshauer_QuantumDecoherence_PhysRep.pdf.

[76] G. Bacciagaluppi, The Role of Decoherence in Quantum Mechanics, Stanford Encyclopedia of Philosophy, 21. April 2020, online zugänglich hier https://plato.stanford.edu/entries/qm-decoherence/

not be explained before. Only its quantitative description was previously overlooked."[77]
Bacciagaluppi writes, among other things: "We describe the overall picture of emergent
structures resulting from decoherence."

I conclude: Obviously, decoherence is happening, thus the loss of the special properties of the quanta, which quantum physics is getting to know better and better—like
indeterminacy, ability to tunnel, particle-wave dualism; decoherence is happening when
transitioning to the macroscopic level, so, that for the same reason, the indeterminacy of
the quanta can not be the cause of chance and coincidence in our classically and relativ-
istically (and thermodynamically) describable macroscopic world. Because the indeterminacy or impossibility to measure location and impulse of the elementary particles at
once already ends on the next higher level of the aggregation of matter. Where exactly
the quantum level ends and the classical macroscopic world begins, seems still seems to
be a topic for discussion.[78] But if I interpret it correctly, single atoms already no longer
show quantum properties when interacting with the environment.

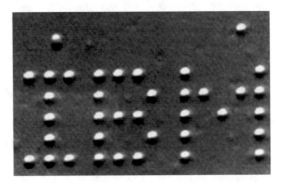

How else could letters ("IBM") and characters (原子 = atom) be made from such isolated atoms in the scanning tunneling microscope, as can be seen in the pictures above
and on the following page?[79] This would make sense to me, because, in an atom, quanta
have already interacted with each other to form the atom. Subsequently, the atom inter-
acts with its environment, whether with other atoms of the same kind, or with the sample

---

[77] Das Originalzitat lautet: "… decoherence is a normal consequence of interacting quantum
mechanical systems. It can hardly be denied to occur – but it cannot explain anything that could
not have been explained before. Remarkable is only its quantitative (realistic) aspect that seems to
have been overlooked for long." H. D. Zeh, What is achieved by decoherence? In: M. Ferrero, A.
van der Merve (Hrsg) New Developments on Fundamental Problems in Quantum Physics (Oviedo
II, Kluver, 1997, S. 441–452; zitiert nach M. Schlosshauer a. a. O.

[78] https://www.scientificamerican.com/article/how-does-the-quantum-world-cross-over/

[79] https://www.researchgate.net/publication/284924836_Materials_characterization_and_the_evo-
lution_of_materials/figures?lo=1   or   https://researcher.watson.ibm.com/researcher/files/us-flinte/
stm11.jpg

surface and the STM probe. This is confirmed by the experimental observation of decoherence, which took place when studying one single Rydberg atom[80] in interaction with only a few photons, which were trapped in a microwave trap, as was demonstrated in the publication cited below.[81] The main author of this article, Serge Haroche,[82] received the Nobel Prize for his work on this topic in 2012.[83] Thus, the experimental demonstration of the phenomenon of decoherence was honored with the Nobel Prize.

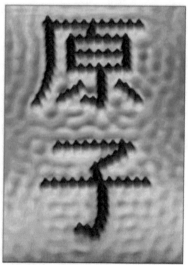

Writing With Atoms. Written literally with atoms, the Japanese Kanji above—each just a few nanometers across—means "atom."

I would like to give the doubting readers another argument to consider: If quantum indeterminacy causes the occurence of chance and coincidence that decoherence would not prevent, then there would be no quantum computers. Well, they don't exist yet in a usable, practical form for any reasonably useful calculation. They exist in small form as research objects that can handle very small, very limited and very special testing tasks. But it would not be theoretically advisable to even research and further develop such new computers if we could really physically reliably attribute the unpredictability of our world to the behavior of the quanta, the elementary particles. We could never achieve any reproducible results; prime number calculations would always deliver different results,

[80] https://www.scientia.global/rydberg-atoms-giants-of-the-atomic-world/.

[81] M. Brune, E. Hagley, J. Dreyer, X. Maître, A. Maali, C. Wunderlich, J. Raimond, S. Haroche, Observing the Progressive Decoherence of the „Meter" in a Quantum Measurement, Phys Rev Lett **77** (24) 4887–4890 (1996)., download the paper here: https://journals.aps.org/prl/pdf/10.1103/PhysRevLett.77.4887.

[82] https://en.wikipedia.org/wiki/Serge_Haroche.

[83] https://physics.aps.org/articles/v5/114

encryption methods would not work. Well, decoherence poses a very serious problem for quantum computers anyway: The qubits must be kept at an extremely low temperature (close to absolute Zero). And even at that, the entanglement of the quanta lasts only fractions of a second, because the subtlest influences from the environment already cancel the quantum state, destroying the quantum state before we got any information. When reading out the results, decoherence takes place, too, the results being a reflection of the behavior of quanta: The qubits, previously have the state 1 or 0, or any state between 0 and 1, all of this with a certain probability. But after reading a result out, only exactly 0 or 1 (and no other state) will show up, and that influences the calculation results which are then only „probably correct" (or probably incorrect). Therefore, calculations of several qubits have to be executed in parallel to increase the probability of a "correct" result. In addition, there is a hardware-related error rate (also the computers known to us have an error rate, so there are error correction procedures, e. g., „forward error correction"). In the presently tiny—compared to the size required for practical applications—quantum computers it is still several magnitudes higher than in conventional computers, namely, still at 10^{-3} (i.e., every thousandth signal is wrong) and, according to the latest publications, for the first time, 10^{-6}, which, however, was achieved under very controlled optimal laboratory conditions. In presently available computers, the rate is 10^{-8} to 10^{-9}, and for 100 Mbit/s data transmission networks, the specification is $<10^{-12}$. Currently available „toy" quantum computers are still lightyears away from this, by a factor of 1 million to 1 billion. If these were the only practical, not to mention theoretical, problems, one might say: Well, that's how a brandnew technology develops. But there are many more and different, even very fundamental hurdles that are anything but certain to be overcome. There are numerous very serious voices saying: We are experiencing a "quantum hype," a bubble that will burst very quickly. The disappointments about the still unachieved goals (which have been considered achievable "in 5–10 years" for 30 years at this point) could become so strong, the confidence in the commercial potential could collapse, a "quantum winter" could set in with up to now billions of dollars in funding drying up.[84] Either way, whether such new computers will become possible and useful or

---

[84] Recent (Feb 23, 2023) publication by a google research team: https://www.nature.com/articles/s41586-022-05434-1. Prediction of a „quantum winter" by Sabine Hossenfelder, a quantum hype critic: https://www.youtube.com/watch?v=CBLVtCYHVO8&t=3s.

A critical voice describing various problems with quantum computers: https://www.quantamagazine.org/why-is-quantum-computing-so-hard-to-explain-20210608/, saying: „The goal in devising an algorithm for a quantum computer is to choreograph a pattern of constructive and destructive interference so that for each wrong answer the contributions to its amplitude cancel each other out, whereas for the right answer the contributions reinforce each other. If, and only if, you can arrange that, you'll see the right answer with a large probability when you look. The tricky part is to do this without knowing the answer in advance, and faster than you could do it with a classical computer. … The problem, in a word, is decoherence, which means unwanted interaction between a quantum computer and its environment — nearby electric fields, warm objects, and other things that can

not, quantum computing, its behavior, and the problems associated with it are a further indisputable evidence that decoherence is a fact. Quanta, with their indeterminacy, which they lose due to even the very smallest interactions with the macroscopic world, cannot be responsible for essential chance and coincidence as we experience it.

And for the first time in the world, a Chinese research institution succeeded in conducting a quantum-encrypted communication.[85] The German science magazine *Spek-*

---

record information about the qubits. This can result in premature "measurement" of the qubits, which collapses them down to classical bits that are either definitely 0 or definitely 1."

Another critic describes the „quantum hype", https://www.technologyreview.com/2022/03/28/1048355/quantum-computing-has-a-hype-problem/ saying: „The most advanced quantum computers today have dozens of decohering (or 'noisy') physical qubits. Building a quantum computer that could crack RSA codes out of such components would require many millions if not billions of qubits. Only tens of thousands of these would be used for computation—so-called logical qubits; the rest would be needed for error correction, compensating for decoherence. ... A decade and more ago, I was often asked when I thought a real quantum computer would be built. (It is interesting that I no longer face this question as quantum-computing hype has apparently convinced people that these systems already exist or are just around the corner). My unequivocal answer was always that I do not know. Predicting the future of technology is impossible—it happens when it happens."

Still another comment about sources of erratic calculations is mentioned here: https://science.thewire.in/the-sciences/quantum-computing-qubits-error-correction-no-cloning-theorem/ „Qubits can maintain superposition only for infinitesimally small intervals of time. Even the slightest interaction with the environment causes a qubit to collapse into a discrete state of either 0 or 1. This is called decoherence. And even before they decohere, random noise caused by non-ideal circuit elements can corrupt the state of the qubits, leading to computing errors."

A more fundamental critical discussion has been published by IEEE Spectrum: https://spectrum.ieee.org/computing/hardware/the-case-against-quantum-computing with the title „The Case Against Quantum Computing", at it says (in between a longer article): „ How is information processed in such a machine? That's done by applying certain kinds of transformations—dubbed 'quantum gates'—that change these parameters in a precise and controlled manner. ... So the number of continuous parameters describing the state of such a useful quantum computer at any given moment must be at least ... 10^{300}. That's a very big number indeed. How big? It is much, much greater than the number of subatomic particles in the observable universe. ... To repeat: A useful quantum computer needs to process a set of continuous parameters that is larger than the number of subatomic particles in the observable universe... How many physical qubits [for error correction] would be required for each logical qubit? No one really knows, but estimates typically range from about 1,000 to 100,000. So the upshot is that a useful quantum computer now needs a million or more qubits. And the number of continuous parameters defining the state of this hypothetical quantum-computing machine—which was already more than astronomical with 1,000 qubits—now becomes even more ludicrous. ... All these problems, as well as a few others I've not mentioned here, raise serious doubts about the future of quantum computing. There is a tremendous gap between the rudimentary but very hard experiments that have been carried out with a few qubits and the extremely developed quantum-computing theory, which relies on manipulating thousands to millions of qubits to calculate anything useful. That gap is not likely to be closed anytime soon."

[85] https://futurezone.at/science/zeilinger-telefoniert-quantenverschluesselt-nach-china/289.001.325

trum published an interview (in German) at the end of 2017 with the leading Chinese quantum physicist Pan JianWei, who studied in China, obtained his doctorate in Austria and then continued his research in Heidelberg.[86] He then returned to China and was a leading figure in the development of a quantum satellite called *Micius,* the world's first satellite of this kind. The quantum-encrypted communication was conducted between Beijing and Vienna. Why was there no noise in the quantum-based encryption, but also understandable communication after decryption, if the quanta only behave so unpredictably and indeterminately? Quite obviously, because there is decoherence.

Nevertheless, the boundary between the level of elementary particles and that of atoms and molecules, which form our macroscopic world, is blurred again and again. So, the unpredictable Brownian particle motion is seen as evidence of the effect of the quanta with its indeterminacy.[87] According to the mentioned source, an experiment can be constructed in which it is decided from the trajectory of a particle in the Brownian motion whether a bomb is detonated or not. This is then underlined with a quote from Nobel laureate Manfred Eigen: "The origin of chance lies in the indeterminacy of these elementary events. [...] Under special conditions, however, there can also be a build-up of elementary processes and thus a macroscopic representation of the indeterminacy of the microscopic dice game. "[88]

What we see with Brownian motion, however, are particles like dust or tiny animals—visible under a light microscope—that are NOT set in motion by quantum particles. Rather, they are set in motion by water molecules! Even atoms are no longer quantum particles, they no longer behave indeterminately, and molecules even less so. That is, it is not the *movements of elementary particles* that push the visible particles, but molecules that no longer behave like quanta, which are so active on the move. And this is because the temperature is, for example, 22°C. Since the *environment-related decoherence* (according to H. D. Zeh) has long since taken place, the water molecules belong to the macroscopic world. They move non-linearly because they are constantly colliding with other water molecules. The movement of tens of thousands of molecules—not just one!—results in an impulse that moves the visible particles more upwards, more downwards, or more to the right or to the left. Nothing miraculously quantum mechanical is going on here!

These relationships are actually well known to physicists. Quantum particles can therefore no longer show their incomprehensible properties in the macroscopic world. There is also neither evidence nor are there even any theoretical indications that the indeterminacy of elementary particles could be the cause of chance in our world. Nevertheless, most physicists insist on this (as they believe) obviously evident explanation, which is then gladly repeated by non-physicists. This reminds me again of the quote from the

---

[86] https://www.spektrum.de/news/ich-bin-froh-ueber-meine-rueckkehr-nach-china/1526607

[87] Cf. e.g. Walter Hehl, loc. cit. p. 47 ff. (Subsection 2.2.5 "From atomic chance to great effect").

[88] M. Eigen, quoted from W. Hehl loc. cit. p. 50.

very readable book by Nobel laureate Kahneman mentioned in the foreword: "Our mind usually works in such a way that we have intuitive feelings and opinions about almost everything we encounter. [...] Regardless of whether we formulate them explicitly or not, we often have answers to questions that we do not fully understand, and we rely on clues that we can neither explain nor defend."[89]

Perhaps it is the lack of evidence that motivates some decoherence theorists to try to describe the transition from the quantum level to the classical macroscopic world in quantum terms. I consider this to be comparable to an equally futile attempt to explain evolution by means of a theoretical description and generalization of the occurrence of mutations. Because evolution does not depend on one mutation alone, nor on several mutations together, but on the new properties developing after these—which cannot be derived from the mutation itself—and the interaction of the organisms with the environment, which also changes. This cannot be described by a *mutation theory*; such a theory could not replace the theory of evolution.

On all levels, starting with the most basic level for which entropy is defined, the one at which substances are formed from atoms and molecules,[90] the chance or coincidence arises solely and precisely because the laws of nature effective on *this* level are non-linear. And because they—tragically, stupidly, pleasantly or fortunately, depending on what happens, and often depending on the point of view or perspective!—also interact with each other. So, the flapping of a butterfly's wing can trigger a tornado a thousand kilometers away or the rustling turning over of a newspaper page in Hamburg can steer the life course of a young chemist 500 km further south onto a completely different, new track. Or not, in fact, in the vast majority of cases, not.

When, a few lines above, using the term *chance/coincidence* and *law* in one breath, I am in full agreement with Eigen/Winkler, who write: "The shape of living beings, the shape of ideas, both have their origin in the *interplay of chance and law.*"[91] And it is just as in the sense of Monod's book titled *Chance and Necessity.*[92] Chances, coincidences happen within the framework and on the basis of physical laws, but not necessarily, not regularly, not predictably—and yet every chance has a cause or, usually, several coinciding causes.

---

[89] Daniel Kahneman (Nobel laureate 2002), "Schnelles Denken, langsames denken", loc. cit. German edition, p. 127.

[90] One (1) atom, one (1) molecule, not even 12 or 15 molecules together, are already a "substance"; one atom is not a liquid, not a solid; 15 atoms or molecules have no volume, exert no pressure, have no surface; such properties, and thus entropy, only occur later; in the next chapter, as an example, I will go into the "metallic property", which also only emerges above a certain minimum number of atoms.

[91] M. Eigen, R. Winkler, loc. cit. p. 87.

[92] J. Monod, loc. cit.

We certainly don't like some coincidences, we hate others, and still others are threatening or even deadly. Then, there are the other coincidences that we love, that we would like to tell our great-grandchildren about when we are old. Coincidences without which everything we know and love would not be as it is. We have no choice but to accept the coincidences of the world and our lives, to take note of them. There is no point in bemoaning the bad coincidences like the Corona virus. It is inappropriate to arrogantly attribute successful outcomes based on fortunate coincidences solely to our own strategically planned achievements. We have no alternative but to sail our limited lives through the turbulent waves of the many non-equilibria of our environment and to stand up to the stormy and constantly changing winds. We can only try to see where they could be of use to us. We should cross patiently, carefully and far-sightedly when they turn against us or even become dangerous.

We just have to get rid of the idea that the world is in equilibrium or should be in equilibrium during our lifetime or during the existence of the solar system. There is no human right to constantly pleasant, beneficial living conditions. Life does not arise in equilibrium and does not exist in such a state. It is in non-equilibrium. A relatively stable non-equilibrium is not the same as *equilibrium*. Life is created by a multitude of dynamic, constantly built up and broken down, constantly transformed dissipative structures. We just have to get used to it. We have to live with the uncertainties, whether we like it or not. The universe will eventually come into equilibrium when entropy reaches its maximum. We don't know exactly what that will look like, maybe there will only be black holes where time stands still and chance and coincidence cease to exist.

Admittedly, these are not particularly inspiring prospects, but we cannot change the increase in entropy on the scale of the universe. Perhaps it will comfort you to know that this is still several hundred or more billion years away. The solar system is likely to experience very large, not particularly attractive changes for Earth a few billion years earlier. But quite independently of that, there is still enough time for each of us, despite our relatively short life expectancy, to enjoy the beauties of the world, the surprises and coincidences that make life so interesting and help us overcome the hardships and *misfortunes* of the non-equilibria.

An unimportant, but simply beautiful event, which is also an example of essential coincidence: Sunrise behind clouds on the horizon, long, quiet waves coming by chance from a suitable direction, to create such impressive splashes on a coincidentally suitable rock ledge, upon which I stand by chance at exactly the right time: Two processes running linearly (sunrise and wave movement) meet in time, while the wave ends non-linearly on the rock and splashes up in part, all of which unexpectedly coincides with my presence on this rock. (a unique photo by the author taken on the coast of Wollongong, Australia, 2004).

What Flows When "Time Flows", and Where Does It Flow to?

7

Abstract

First, we look at what different physicists think about time: Is it an illusion? Does the timelessness of the quanta mean that there is no time at all? Can the arrow of time be reversed? Do we live in one of many universes? We then think about what the formulation "time flows" could mean. It is explained that time cannot *flow* and cannot *pass*. Finally, a new hypothesis is presented that describes time as an emergent phenomenon through the flow of entropy. This new hypothesis is experimentally verifiable.

... *Black holes in which time stands still*—how do I come up with such a formulation at the end of the previous chapter? How can time stand still, is that even possible? What is it, time? What is the nature of time? What is it made of, if you can ask that question, what is it that is flowing when we say "time is flowing"?

These questions are still an unsolved mystery for most of the people, natural scientists, philosophers and many others, who occasionally think about them. John Wheeler (a prominent theoretical physicist)[1] once wrote: "Explain time? Not without explaining existence! Explain existence? Not without explaining time! Reveal the deeper hidden connection between time and existence [...] A task for the future."[2] With this book, and especially this chapter, I would like to make an attempt to contribute to the accomplishment of this task.

---

[1] https://de.wikipedia.org/wiki/John_Archibald_Wheeler.

[2] J. A. Wheeler, American Scientist 74 (1986), pp. 366–375, quoted from Carlo Rovelli, "The Order of Time", Rowohlt 2018, p. 100/101 or the footnote 72 there.

Many wise people have expressed deep thoughts about the nature of time. For Aristotle (Book IV), time is "a number of motions with respect to 'before' and 'after'."[3] Time points ("nows") are not units of time, but boundaries of time intervals, the length of which is measured by the heavenly bodies.[4] According to Isaac Newton, "absolute time (which is nothing other than duration or the perseverance of the existence of things) remains the same, whether the motions be be swift, slow, or null."—"Absolute, true, and mathematical time, from its own nature, passes equally without relation to anything external, and thus without reference to any change or way of measuring of time (e.g., the hour, day, month, or year)." (cited here as paraphrased in Stanford Encyclopedia of Philosophy)[5]

Let's jump into modern times. Henning Genz suggests choosing the radius of the universe as "the representative of the 'time' parameter."[6] For Genz, time is connected with movement, with the expansion of the cosmos. I have a problem with that, because, as long as the gravitational constant is actually constant, the universe expands, but not the earth—nevertheless, we observe time here on earth as a real phenomenon, the nature of which we want to understand.

In an introductory article *What is time?*, Andreas Müller discusses the concept of time (in German).[7] The author makes it clear that, even after centuries of reflection, there is still no physical explanation of time. Strangely, entropy is misrepresented here, in that it is described as *only and always increasing*. This, as we have discussed in previous chapters, is the case on the scale of the entire universe, but not in all the open subsystems in the world that are far from equilibrium—that is, the current majority of all parts of the universe. Finally, the author sums up: "The physical concept of time must be further explored." I do not share his "deep hope" for enlightenment in "new, higher-level theories, such as quantum gravity" and, in particular, in *Loop Quantum Gravity*. After the author sets his hopes on these theories, he continues sadly (because these concepts are "relatively unenlightening"): "We pay with the clarity and win the price of knowledge." It should be said that *Loop Quantum Gravity* and other theories of this kind have not yet yielded any knowledge, but have already become unenlightening at this point. I would like to start an attempt here to refute this and to make it possible to gain knowledge

---

[3] https://en.wikipedia.org/wiki/Physics_(Aristotle)

[4] https://ndpr.nd.edu/reviews/time-for-aristotle/

[5] https://plato.stanford.edu/entries/newton-stm/

[6] Henning Genz, "Wie die Zeit in die Welt kam—Die Entstehung einer Illusion aus Ordnung und Chaos", Rowohlt Taschenbuch 2002, p. 228 ff. (English introduction into a new German edition: "How time came into the world—The origin of an illusion from order and chaos", https://www.hanser-literaturverlage.de/en/buch/how-time-came-into-the-world/978-3-446-18742-9/)

[7] https://www.spektrum.de/astrowissen/zeit.html. Adam Becker is also promoting the "loop quantum theory" as a tool for better understanding of space and time, which as well does not contribute any deeper understanding but abstract theory, detached from reality, without any tangible (and especially not provable, falsifiable) results, cf "What Is Spacetime Really Made Of?" Scientific American, Febr 1, 2022 (https://www.scientificamerican.com/article/what-is-spacetimereallymade-of/)

without losing clarity. In the following, I will develop a hypothesis of how we might understand the nature of time.

Let's start with what we know about time from our everyday lives: It inexorably marches on, we never experience it running backwards. That's called the *arrow of time*. The arrow of time is unmistakable, we all have to deal with it every day. But the basic theories with which we describe nature do not know *time*, at least, time does not have a direction in them. This applies to classical physics such as Newton's theory of gravitation, as well as to Einstein's theory of relativity and to quantum mechanics. For the results of calculations within the framework of these theories, it is irrelevant whether the time has a positive or negative direction. This tempts some physicists and, as a result, philosophers to declare that *time* is an illusion and to believe that we can explain and describe the world without *time*. I will come back to this in more detail later. Ilya Prigogine called the contradiction between our daily experiences and the *timelessness* of the basic laws of nature *the paradox of time*.[8] His fundamental book, written together with Isabelle Stengers, convincingly demonstrates why there is really an arrow of time, and thus also time: First, because of the irreversibility of processes in non-equilibrium, in which we live, and thus because of entropy; secondly, because of the existence of "events", i.e., processes that happen suddenly. We talked about this in the previous chapters and examined the causes of chance and coincidence, which are veritable events. Because events happen suddenly, they cannot be derived from linear deterministic laws, because these do not provide for *suddenness*. Thirdly, because certain events can have a major impact on the course of the world and, for example, also on evolution.[9]

Leading scientists (such as Prigogine, Barrows, Penrose, Hawking[10]) agree that the arrow of time is directed from the past to the future due to the second law of thermodynamics. Jacques Monod writes about this in his book:[11] According to this, the irreversibility of evolution can be considered as an expression of the 2nd law of thermodynamics. If I may interpret this for you, it means: The evolution develops without a goal, but with continuously new results, which is associated with entropy production and makes the individual evolutionary steps irreversible. And the irreversibility clearly shows us the direction of time, from the past through the present into the future.

---

[8] Ilya Prigogine, Isabelle Stengers, "Das Paradox der Zeit—Zeit, Chaos und Quanten", Piper-Verlag 1993 (the English edition "The end of certainty: time's flow and the laws of nature" was published in 1997 by Simon & Schuster NY)

[9] I. Prigogine, I. Stengers, loc. cit., p. 74/75.

[10] Stephen W. Hawking: "Eine kurze Geschichte der Zeit", Rowohlt 1988, original english edition "A Brief History of Time",   Bantam Dell Publishing Group, 1988; a pdf is available at docs.google.com: http://bit.ly/3E5ST32

[11] "The evolution in the living nature is thus a necessarily irreversible process, determined by a direction in time; the direction is the same as that prescribed by the law of increasing entropy, that is: by the Second Law of Thermodynamics." translated by the author, cited after the German edition "Zufall und Notwendigkeit": J. Monod, loc. cit. p. 113.

Had Prigogine and Stengers explained what *the nature of time* is in their book, this chapter of my book would not have to exist, I would only quote their book. But the two authors neither commented on this nor was it their goal. We therefore want to try to find out in the following whether we can learn to understand the nature, the essence of time.

Is Time an Illusion?

More than just a few physicists and philosophers think that time is an illusion. Henning Genz writes about this in the subtitle of his book *How time came into the world:* "The emergence of an illusion from order and chaos". He explains it in the text as follows: "As an emergent property of large systems, time is indeed real, but still [...] an illusion [...] a real illusion."[12] Lee Smolin writes, in his book *Im Universum der Zeit* (original title: *Time Reborn*)[13]: "The idea that time is an illusion has the status of a philosophical and religious truism." The laws of gravitation, whether Newton's or Einstein's, yield the same results regardless of the direction of the arrow of time. These and many other physical laws are basically *timeless,* in the sense that the direction of time does not matter. This motivates some physicists to believe that these laws also allow for the reversal of the arrow of time.

 H. Genz proposes a concept worth considering. According to his ideas, we should not speak of an *arrow of time,* but of an *arrow of processes.*[14] If he had not made the restriction that all processes are running from higher to lower order—that is, processes that lead to higher entropy—we would already be a good step further. Because we also have to ask ourselves: And what about processes that we are surrounded by in manifold ways, that lead to systems of higher order, which are thus associated with a reduction in entropy, and thus with entropy export? Where does the *arrow of processes* point in these systems? We also have to ask this question of the physicist H. Dieter Zeh, whom we have already met as the creator of the *decoherence theory.* For him, in his book on the physics of time, it is clear: "The thermodynamic direction of time is characterized by the increase in entropy."[15] And we would have to ask the same question of Stephen Hawking (or Walter Hehl, who quoted him), because of the claim: "The increase of disorder or entropy is what distinguishes the past from the future; this gives the direction of time."[16] Hehl continues: "In the macroscopic world of our everyday life there is only

---

[12] Henning Genz, loc.cit., p. 41.

[13] Lee Smolin, "Im Universum der Zeit" Pantheon-Verlag 2015 (original edition 2013 "Time Reborn", Houghton Mifflin Harcourt, N. Y.).

[14] H. Genz, loc. cit. p. 188.

[15] H. D. Zeh, "The Physical Basis of The Direction of Time", 5th edition Springer 2007 (https://link.springer.com/book/10.1007/978-3-540-68001-7), cited after original German edition "Die Physik der Zeitrichtung", Springer 1984, p. 23.

[16] quoted in Walter Hehl loc. cit. (English edition), p. 55.

one direction: Entropy increases. A reduction is only possible artificially and with added energy."[17] This is not correct, as we have learned, because, in our macroscopic world, there is a constant reduction in entropy in all possible open subsystems, not only by artificial means, but quite naturally. For billions of years. The entropy increases on the scale of the universe, with the growth of black holes and the increase in their number, but not on Earth or in the Sun alone. Then, does the direction of time change, does time go back when entropy decreases dramatically in all the non-equilibrium systems around us and in our own bodies because it is exported (at least before we die)? Certainly not.

From our everyday experience, it is inconceivable that, for example, a tree that has been toppled over and partly broken during a storm could stand up again spontaneously and, a few weeks later, cast shadows again with its intensively greening leaf canopy. Or maybe if the storm came with exactly the same strength from exactly the opposite direction? Of course not, even then. Nevertheless, there are far more than just one or two physicists who think that this might be hard to imagine, and there is allegedly certainly a very small probability that something like this could happen, but if we only waited long enough, maybe longer than the time the universe has been in existence, something like this could happen.[18] At least, however, we could note that, with this argument, *time* is seen as reality.

In the article cited in the footnote below,[19] it is claimed that there are more and more indications that time is an illusion, that it does not exist at all. If we follow the argumentation of this article, time was not created at the Big Bang, and there was no time before that. Craig Callender argues, in an article entitled *Is time an illusion* in *Scientific American,* that everything we express today in units of time (days, hours, minutes and seconds)

---

[17] Walter Hehl loc. cit., p. 57.

[18] http://www.bbc.com/earth/story/20160708-the-past-is-not-set-in-stone-so-we-may-be-able-to-change-it The analogy used here was not my toppled tree, but a less conceivable one: Two gases are mixed in two containers, in both the same mixture of nitrogen and carbon dioxide. Now we connect the two vessels with a pipe. Is it conceivable that the two gases mix spontaneously and "left" only nitrogen is found, "right" only carbon dioxide? Thermodynamics says "no", that is not possible. The defenders of the hypothesis "time is an illusion, and there is no arrow of time" say: "Yes, the gases can unmix spontaneously, it is extremely unlikely and it may take extremely long, maybe longer than the universe is old, but it is possible." It is overlooked that the mixture of the two gases in the two now connected vessels corresponds to an equilibrium state, that is, a maximum of entropy. So we would have to bring an extremely high amount of energy to separate the gases so that the entropy can be extremely reduced. This cannot happen spontaneously, not by itself, not in more than 15 billion years. Other authors like H. Genz in "How time came into the world", loc. cit. p. 188, use similar arguments like separating gas mixtures, simple pendulums or gears that could make something extremely unlikely based on probabilistic considerations. Such considerations are, because they are based on statistics, ultimately based on equilibrium behavior. From this no conclusions can be drawn for complex systems that are far from equilibrium.

[19] Zeeya Merali and Chad Hagen, "In Search of Time's' origin. What happened before the Big bang?" http://nautil.us/issue/9/time/in-search-of-times-origin.

could also be described in other units.[20] This would be comparable to the idea that we do not necessarily need money: Instead of paying 100 euros for a pair of shoes, we could also exchange the shoes for 50 cups of coffee. Similarly, we could say "Every rotation of the Earth corresponds to 108,000 heartbeats" instead of "75 beats per minute" and "the Earth rotates once a day around its axis". I would—with all due respect—clearly reject this, because the heart rate is not regular at all, neither from person to person nor for a single person: because it depends on constitution, actual situation, state of health and much more.

But even if heartbeats were exactly regular and the same for all people, or if we were to use a metronome instead of *time-currency heartbeat* and set it to 75 beats per min: What would that be other than a clock with which we measure time today? Then, a rotation of the Earth would correspond to 108,000 metronome beats. Yes, that's how time is measured, so what did Craig Callender want to tell us?

In another article, the claim is even made that the assumption that "time is an illusion" is gaining more and more acceptance.[21] I would like to quote this word for word in excerpts (translated by myself): "This flow of time is very familiar to us and at the same time extremely mysterious—but nevertheless probably a sheer illusion. For more and more physicists and philosophers are coming to the conclusion that time objectively does not exist at all. 'Recognizing this may be the greatest intellectual challenge humanity has ever faced,' says philosopher and physicist Vesselin Petkov of Concordia University in Montreal, Canada." The article ends with a quote from a very respected theoretical physicist: "… Carlo Rovelli, professor of physics at the University of Marseille, also considers time to be an illusion: 'Even though I can't prove it, I'm convinced that time doesn't exist. I believe that there is a way to describe how nature works without using the concepts of time and space. 'Space' and 'Time' will only remain meaningful within certain approximations—just as the concept 'water surface' loses its meaning when we look

---

[20] cited (and translated) after Craig Callender in the German popular science magazine Spektrum der Wissenschaft, 10.2010, p. 33 (https://www.spektrum.de/magazin/ist-zeit-eine-illusion/1044180); see also Craig Callender, in: Scientific American (June 1, 2010), https://www.scientificamerican.com/article/is-time-an-illusion/. In the following article, some physicists from the USA and GB were interviewed and asked for their opinion on the question "What is time?", their answers essentially: "an illusion" https://www.space.com/29859-the-illusion-of-time.html.

[21] Rüdiger Vaas "Gestern und Morgen sind eins", ("Yesterday and tomorrow are one") https://www.wissenschaft.de/umwelt-natur/gestern-und-morgen-sind-eins/ (18. 02. 2008, German). My answer to Carlo Rovelli is supported by Christian Wüthrich, a philosopher of physics at the University of Geneva. He said, when considering a liquid: "Ultimately it's elementary particles, like electrons and protons and neutrons or, even more fundamental, quarks and leptons. Do quarks and leptons have liquid properties? That just doesn't make sense, right?... Nevertheless, when these fundamental particles come together in sufficient numbers and show a certain behavior together, collective behavior, then they will act in a way that is like a liquid." cf Adam Becker's article cited in https://www.scientificamerican.com/article/what-is-spacetime-really-made-of/. Hence Wüthrich supports the concept of emergence.

at water molecules in detail: If we look closely enough, there is no such thing as a water surface. It is quite similar with time and space: 'They are only macroscopic approximations—illusions that our consciousness has created in order to understand reality.'"

Once again, this shows the problem already discussed in the previous chapter, according to which, all too often and too readily, the fact is overlooked or denied that new laws arise on higher aggregation levels of matter. Of course there is a water surface, and yes, it consists of individual water molecules, but in the moment when these molecules interact with each other and change from the gaseous to the liquid state and then form a boundary surface to air—the *water surface*—new phenomena arise, in this case, e. g., the surface tension. A single water molecule does not have a surface tension, just as water in the gaseous state does not. But in the liquid state and at the boundary surface to air (or to other media such as glass or fat), the surface tension exists as a new phenomenon. A phenomenon that we can measure and observe the effect of, for example, when water striders (Gerridae) walk on the surface of a pond and do not sink. You can read more about Carlo Rovelli's latest book, in which he sets out his understanding of time, a little later on.

According to Lee Smolin, the *illusion* argument arises from what he calls a *cosmological fallacy:*[22] This would be—as one can see with Newton's law of gravitation—the premature conclusion that, e.g., Einstein's theory of relativity, which applies to some or even all subsystems, also applies to the cosmos as a whole: "The universe is a different kind of entity than any of its parts. Nor is it simply the sum of its parts."[23] We can at least agree with Smolin's the last sentence, because we have discussed in detail in the previous chapters the fact that, on a higher aggregation level, *additional* new laws occur.

Let's jump back to Prigogine/Stengers and Monod with their justification for the existence of the arrow of time. I want to ask about it now in a different way: What do we observe around us while we percieve that time flows from the past through the present into the future? The answer: We observe *changes* around us, as well as in and on ourselves. We observe that the plants grow, bloom, develop fruits and die. We observe clouds that come and go, arise and disappear. We observe the rain that begins and ends. The sun rises, apparently moves from east to west, and sets again. Spring, summer, autumn and winter come and go. Geologists, archaeologists and paleobiologists discover traces from even the most distant past of continental drift, climate change and the evolution on our planet. Astronomers look into the distant past when they examine objects millions and billions of light years away. All this shows us the course of time and the arrow of time.

We have discussed in the previous chapters why there is emergence, becoming, ripening and passing away in the world, with all the components that exist in the universe. Nothing has always been there, everything has developed from a beginning; nothing is

---

[22] Lee Smolin, loc.cit., p. 147.
[23] Ibid., p. 149.

constant, everything decays. Not only the plants and animals and we ourselves, but also the mountains, the continental plates, the sun, the stars and the galaxies.

While time, according to our current understanding, is flowing from the past into the future, *bifurcations* arise for all possible processes and systems in the world, and also for ourselves. So, one system went in one direction, the other in another one; we decided to take this path at the fork in the road, and not the other one. Thus, history can not be undone, it can not start all over again along a different course at the fork in the road.

Can the Direction of Time be Reversed?

This is often where the imagination of some physicists and science journalists starts to bloom, and not just in popular science magazines and TV documentaries. For example, it is claimed that two mixed gases could spontaneously demix, even though it would take an incredibly long time. Or, as Sean Carroll wrote in *Scientific American*, "milk molecules" (which don't exist as such, milk is an emulsion consisting of a lot of different molecules) could maybe somehow and at some point spontaneously separate from coffee after the two have been mixed together, however: "Although it is physically possible for all the milk molecules to spontaneously conspire to arrange themselves next to one another, it is statistically very unlikely. If you waited for it to happen of its own accord as molecules randomly reshuffled, you would typically have to wait much longer than the current age of the observable universe."[24] The justification sometimes given is quantum physics, in which *retrocausality*[25] is said to be possible. It is also often said (e.g., by C. Rovelli) that the arrow of time, as we experience it, is merely *psychological*. The *thermodynamic* arrow of time need not be in harmony with the *psychological* one[26], it just seems that way because our memory works that way. The defenders of this theory like to present a series of complicated arguments as to how memory is effectively playing a trick on us, and how what we actually know as the future is interpreted as the past. Apart from the fact that these are all thought experiments, the whole argument has a serious flaw: The thermodynamic arrow of time is determined by the production of entropy in non-equilibrium processes. The retrocausality capability on the quantum level is lost due to decoherence during the transition to the higher aggregation levels of matter, that of atoms and substances, as atoms can no longer tunnel, even though electrons do so constantly. On the quantum level, there is no thermodynamic arrow of time determined by entropy. Because, on the quantum level, there is no entropy. Entropy only exists on the level of substances,

---

[24] https://www.scientificamerican.com/article/the-cosmic-origins-of-times-arrow/

[25] https://en.wikipedia.org/wiki/Retrocausality

[26] cf. this overview about "The arrow of time", https://www.exactlywhatistime.com/physics-of-time/the-arrow-of-time/, and also subsection "Psychological/perceptual arrow of time" in https://en.wikipedia.org/wiki/Arrow_of_time

of matter made from atoms and molecules, according to its definition.[27] Metabolism takes place in our memory, thus entropy is produced there. Therefore, the psychological arrow of time in our heads has the same direction as the thermodynamic one.

In some articles, the example of a coffee cup falling off the table and breaking on the floor, spilling coffee on the carpet, is lovingly used. If we only had enough time, we might be able to observe, at some point, that the cup spontaneously reassembles itself and the coffee is back in the cup (although it would probably be quite cold by then). It is admittedly extremely unlikely, as one can read, but nevertheless not impossible. These arguments lack any reasonable basis. They are sheer nonsense. After the cup breaks and the coffee spills, the entropy of this subsystem is greater than before because there is more disorder. And if we just waited, no supercritical energy would be introduced into the system *broken coffee cup with spilled coffee* from the outside, so that, without very specific and high energy input and entropy export, the entropy could not be reduced to the lower initial state. This cannot happen spontaneously without energy input. Moreover, the water would have evaporated after a day, the water molecules we need for the coffee would be irretrievably distributed in the atmosphere, connected with an increase in entropy there. So, this speculation is in stark contrast to entropy and its properties.

The fantasy arguments described above are based on the fact that Newton's and Einstein's laws of gravity and many other fundamental physical laws are neutral with respect to the direction of time. This means that, regardless of whether time runs forwards or backwards in the equations, the results are the same. Therefore, it is assumed that time does not exist as such. Physicists feel obliged to describe a universe without a direction of time because Einstein's theory has been confirmed so wonderfully and precisely, not least with the discovery of gravitational waves, which Einstein considered existent but unmeasurable.

Jacques Monod probably did not know Prigogine's work (Prigogine, on the other hand, knew Monod's); I can find no other explanation for why Monod allowed himself to make a strange statement in his otherwise very clear book: While he sees the arrow of time in the second law of thermodynamics—*constant increase of entropy in the world*—in the sense of Prigogine, he does not seem to have understood it. Because, he writes, the sentence allows that the entropy in a "macroscopic system in a change of very small range and for a very short period of time" can decrease.[28] This means that such a system could "somehow go back in time". He therefore suspected a "mechanism of time reversal" in evolution. This is (with all due respect to Monod) absurd.

---

[27] more precisely: when heat is generated or consumed, and a "temperature" can be measured or could be; entropy is defined by the equation $dS = dQ/T$, or in the 2nd law of thermodynamics $dH = dG - TdS$, thus $dS = (dG - dH)/T$, where H is the enthalpy of reaction, G the free energy of a reaction, thus terms for phenomena which only occur when substances can react with each other, and T (temperature) only when particles are moving and molecules are vibrating resulting in a "temperature".

[28] Jacques Monod, loc. cit. p. 114.

Oddly enough, such speculation does not seem to be unique, nor does it seem (in view of the fifty years since then) to be considered outdated. Because Carlo Rovelli formulated something similar when he wrote: "The entropy increases. Its growth is what we experience as time. I'm not sure if this representation is plausible, but I don't have a better one."[29] When reading and digesting this sentence, you should remember his quote a few pages earlier, which shows that he believes that time does not exist. In his opinion, we experience something like a progression of time, but that experience only occurs because we observe change and conclude from it (experience) that time has passed.

The Timelessness of the Quantum—So Time is an Illusion After All?

In his book, Rovelli sets out his understanding of time. It develops from his work as a theoretical physicist, particularly with theories of *loop quantum gravity*. From the assumption (which I agree with) that there is no time on the quantum level,[30] he concludes that—because the quanta form the atoms, and these, in turn, the substances, and thus the universe and life on Earth—there is also no time on the level of substances and life. With all due respect, Rovelli makes the mistake that I have already mentioned several times in this book: He overlooks that, on higher aggregation levels of matter, new properties and laws arise. So, on the quantum level, there is (in contrast to Rovelli's arguments) also no entropy, a measurable physical property that plays an important role in Rovelli's arguments; it arises as an emergent property on the level of substances, matter, i.e., on levels above the quantum level. Rovelli also completely overlooks that— while entropy increases on the scale of the universe—it decreases very sharply in non-equilibrium systems. So, his statement quoted above, according to which the *growth* of entropy is what we experience as time,[31] is a great misunderstanding of entropy and

---

[29] Carlo Rovelli, "The Order of Time", Rowohlt 2018, p. 130.

[30] "On the most fundamental level we know today [the quantum level], there is thus little that resembles time in our experience world. There is no special 'time' variable, no difference between past and future, and no space-time." C. Rovelli, loc. cit., p. 160 (cited after the German edition and translated by the author). I have commented critically on this and other statements in Rovelli's book in a letter to the editor of a review of his book at spektrum.de, see "Reader Opinion" (in German) under https://www.spektrum.de/rezension/buchkritik-zu-die-ordnung-der-zeit/1644578. The "Loop Quantum Gravity" is also addressed in Adam Becker's article "What Is Spacetime Really Made Of?", Scientific American, Febr 1, 2022, https://www.scientificamerican.com/article/what-is-spacetime-really-made-of/

[31] An ultimately untenable thought that we have already read and discussed before with H. Genz and H. D. Zeh, which, consistently thought through, would mean: In systems in which entropy is partly very strongly reduced (i.e. exported) due to their non-equilibrium character, the direction of time reverses. This is absurd.

our experiences. Because where we *experience* something, *entropy decreases* constantly. We *pump* enormous amounts of energy into our brains, we send entropy into the environment in the form of our excretions, and, in addition, we radiate it as body heat, especially via the head. Moreover, we cannot directly *perceive* entropy. (More about personal time perception in the next chapter.)

Rovelli expresses another—in my opinion, very strange—idea: We humans have a "fuzzy view" of the universe, we describe the world quite inaccurately. I agree with that. But then he says: "Consequently, the difference between past and future ultimately depends on this fuzzy view. If I could take into account all the details, the exact state of the world at the microscopic level, would the characteristic aspects of the course of time disappear? By all means, yes. As soon as I look at the microscopic state of things, the difference between past and future disappears. For example, the future of the world is no more and no less determined by the present state than the past."[32] This is the view of a quantum theorist, starting with the actual timelessness of the quanta, which is not applicable to micro- and macroscopic systems.

Based on this, he develops his understanding of time in the macroscopic world, which therefore does not exist, but is only experienced by us. Namely, approximately as we also believe that the sky revolves around us, around our earth, which we have already determined is not the case, even if it took thousands of years to learn this.[33] The *time* as we know it, he says, is therefore only a question of our perspective.[34]

The fundamental error that Rovelli makes, in my opinion, is twofold. First, we cannot derive the properties and laws of the macroscopic world from a knowledge of the quantum world, no matter how accurate it is. This means, inter alia, that the non-existence of time on the quantum level is no indication of whether it exists on a higher aggregation level of matter or not. Secondly, the entropy increases on the scale of the universe, but not in open non-equilibrium systems, where it decreases. And Rovelli has overlooked that, just as time does not exist on the quantum level, neither does entropy, but it does in our macroscopic world.

We must now again, and not only because of Rovelli's arguments, deal with the problem of the quantum level. The theoretical physicist H. Dieter Zeh, whom we have met in the previous chapter as the founder of the decoherence theory, was also concerned with the problem of *time*. Strangely enough, in his considerations of time, he starts from the

---

[32] C. Rovelli, a. a. O, p. 33. In the footnote 20 mentioned there, he practically shows a deterministic view of the world, with which I do not conform at all, it should have become clear from the previous chapters: Non-equilibrium systems are not deterministic, they only show "deterministic chaos", which is by no means the same as determinism, and even deterministic chaos is not often found in the chaos of this world, from which spontaneity (and chance!) arises.

[33] C. Rovelli, loc. cit., p. 123.

[34] C. Rovelli, loc. cit., p. 161.

quantum level.[35] Strange, because Zeh, as the discoverer of the decoherence phenomenon, should have taken decoherence into account! He leads us to the theory of quantum gravity, for which the *Wheeler-DeWitt equation* is the basis and does not describe any time dependence. This is, according to my understanding, in line with the timelessness of quanta. But he writes further: "This common consequence of the absence of an absolute time and quantum theory thus requires a much more extensive timelessness as in the theory of Barbour and Bertotti [...]. " He refers to this equation as "time-independent wave function over his [the universe, author's note] entire state space, so that one can say that the wave function takes the place of time." What I find strange about these considerations is this: H. D. Zeh was the first to discover and describe decoherence, he has seen that they were experimentally confirmed. Why did he not also deduce from his understanding of the decoherence that the timelessness of quanta, like all their other properties, no longer has to apply in the macroscopic world, and that entirely new properties arise there? It seems to be widespread, the habit of overlooking this fact of the emergence of new properties at a higher organizational level of matter.

Monod had already written in the foreword to his book that the complex structures and functions of organisms can not be derived from the theory of the genetic code, they can not be analyzed even at the molecular level: "One can neither predict nor explain all the details of chemistry with the help of quantum theory, although hardly anyone doubts that this theory forms the universal basis." [36]

At this point, we should take a critical look at the older book *The End of Time*[37] by British physicist Julian Barbour. With his theory, he wants to remove *time* from physics, time is said to be a *redundant* (unnecessary, superfluous) concept. For him, time is essentially a sequence of images, snapshots that merge into each other: "I look at you, you nod. Without this change, we would not perceive time."[38] And further: "Change is real, but time is not real. Time is just a reflection of change."[39]

First, I agree, because "change is real" expresses the fact that becoming and passing away are real, and these processes are connected with entropy production. However, I disagree with the statement that, without perceived change, time could not be perceived.

---

[35] see "Time in nature" (German) https://www.thp.uni-koeln.de/gravitation/zeh/zeit-in-natur.pdf and "What does it mean: there is no time?" (German) https://www.thp.uni-koeln.de/gravitation/zeh/Keine_Zeit.html. For an overview about H. D. Zeh's papers, see https://www.thp.uni-koeln.de/gravitation/zeh/index.html

[36] J. Monod, loc. cit. p. 19/29.

[37] Julian Barbour, "The End of Time", Oxford University Press 2001; and "The Nature of Time" https://arxiv.org/pdf/0903.3489.pdf, 2009.

[38] in: https://www.space.com/29859-the-illusion-of-time.html

[39] This puts him on Mach's shoulders, whom he quotes in "The Nature of Time": "It is utterly beyond our power to measure change of things by time … time is an abstraction at which we arrive by means of the changes of things … "

And not just because of the fact that the question of our perception is not decisive for the question of whether time exists and, if so, what the nature of time is. Another weakness of his theory is that Barbour's *changes* are practically only "movements of objects".[40] Barbour develops his hypothesis, according to which time is redundant, in a detailed theoretical article.[41] Finally, he comes to this conclusion: "The flow of time and motions are illusions." And it culminates in the following basic statement: "Time as an independent concept has no place in physics."

Do We Live in One of Many Universes?

Julian Barbour promises us another theory of time with his newer book *The Janus Point—A new theory of time*.[42] When I found out that the book was going to be published, I immediately placed a pre-order with my bookshop. I was very disappointed after reading it. At least the author no longer sees time as *physically redundant,* in contrast to his earlier statements and theories. He, too, now believes that the growth of entropy determines the arrow of time, and thus that time is real. But then fantasy takes over. Based on the uncertainty of why the universe is not symmetrical in every respect—because the arrow of time is associated with a symmetry break—he develops the hypothesis[43] that, in order to maintain symmetry, there should exist a *mirror universe*, in which the arrow of time points in the opposite direction. The *advantage* of this fantastic assumption is that it is, in principle, neither verifiable nor falsifiable, just like the theories of multiverses popular with many physicists, according to which there are many other universes outside of our own. Advantage, because Barbour and his colleagues can continue to play with such ideas, but it's not real science, as such a "hypothesis" is principally not falsifiable: We don't have any access to information from such a mirror universe. There are numerous variants of these hypotheses, one of which is presented to us by Sean M. Carroll, who maintains a close exchange with Julian Barbour.

---

[40] Barbour's understanding of "changes" is limited exclusively to the position of objects and their movements in space, he defines the "time" here on the basis of Newton's theory with regard to movements of masses and changes in potential energy; the term "entropy" is not mentioned even once, although "change" in a complete understanding is much more than just movement!

[41] Julian Barbour, "The Nature of Time" https://arxiv.org/pdf/0903.3489.pdf, 2009.

[42] Julian Barbour, "The Janus Point—a new theory of time", Penguin 2020. A simple summary can be found here (in English): http://www.januscosmologicalmodel.com/januspoint.

[43] But what Barbour immediately calls "theory", is not appropriate from a scientific-theoretical point of view; one could better say "hypothesis" or "presumption" or "guess".

He assumes, in an article, that the basic physical laws and the standard cosmological model are all symmetrical with respect to time.[44] He does not claim that time is an illusion, but that the arrow of time contradicts the best-confirmed laws: the laws of microphysics and gravitation do not distinguish between past and future. He continues: "The arrow of time is nothing other than the tendency of systems to develop into one of the numerous systems with high entropy." First of all, it is true that entropy creates the arrow of time. But by now, we have learned that all the systems around us that we know and live in (including ourselves) are systems in which entropy is often decreasing (when and if new and more dissipative structure are formed), and entropy is ultimately exported somewhere into the cosmos. I know that I am repeating myself here: If the arrow of time is nothing but the *tendency towards high entropy*, then, in all systems (including, for example, myself) in which entropy decreases, is the arrow of time opposite to the overall universe? No, definitely not.

He also advocates this strange view (one that I already mentioned) that a spontaneous unmixing of milk and coffee, once they are nicely mixed together, is indeed unlikely, but nevertheless not impossible. This would, following his logic about the arrow of time being parallel with entropy *increase*, mean that the *time arrow could spontaneously reverse itself* at some point, and this without any external effort, just on its own; not likely, he says, but still possible. But that is simply impossible, and this because of entropy. When the milk was mixed with the coffee, the entropy increased, visible in the *disorder* of the tiny milk and coffee components that are now just statistically evenly distributed. To reduce entropy—that is, to restore a certain order, milk in one area of the coffee cup, for example, at the top, coffee below in all other areas—*energy must be expended*, and not just a small amount. And this even quite specifically, because cooking or cooling does not help. We would have to look for and separate the milk nanodroplets, an extremely laborious task, to say the least, but, in fact, definitely impossible. Only with supercritical energy can entropy be reduced in open systems—here, now, the coffee cup filled with coffee/milk, which is obviously very popular with theoretical physicists. But even with that, the time arrow would not reverse its direction! With the reduction of entropy, time is not turned back! Energy input, entropy export, structure building, all of this happens over time, taking some time, and then time moves on, in the same direction.

Carroll wonders why the entropy was so low at the beginning of the universe, because it must have been low, at least, lower than today, as it is constantly increasing. After all, the universe *started out ordered* (= low entropy) and then became *more and more disordered* (= high entropy): "cool, diluted, lumpy". But he finds that "very unnatural in view of the fact that states of low entropy are so rare". *Fact? Rare?* On the one hand, such states, when we look at our near and far and very far environment—the universe itself,

---

[44] S. Carroll, "The Cosmic Origins of Time's Arrow", Scientific American June 1, 2008, https://www.scientificamerican.com/article/the-cosmic-origins-of-times-arrow/

after all, is also highly structured,[45,46] cf. Chap. 6—, are anything but rare. Otherwise, all these structures and highly structured systems (= low entropy) that we can observe would not exist, and all of these are extremely far from an equilibrium state (which would exhibit maximum entropy). On the other hand, one must realize how and when entropy arises: It is associated with (or is a by-product of) energy and matter turnover. When the universe began, when *energy* and *matter* first came into existence at all, there could be *no previous turnover*. So, the entropy was an even zero, more precisely: it was non-existent. And only *then*, with the emergence of entropy, did time begin to exist.

But Carroll does not seem to have taken this into account, and he continues with the view that the universe could have started just as we imagine the end: With a contraction of an extremely far-flung universe to an infinitely small point, that is, exactly the reverse of what is generally assumed today as being the beginning of the universe: the "big bang." According to Carroll, this is permitted by the fundamental physical laws (gravity, quantum mechanics, the standard model of cosmology) because they are all timeless and do not contain an arrow of time.

According to Carroll, we only observe an arrow of time in the direction that we know because, somewhere outside our universe, another universe exists in which time runs backwards; we are therefore only part of a system of *multiverses,* in which, when taken together, the symmetry of the arrow of time is fortunately preserved (by the way: if that were the case, there would have to be a way for the different universes to interact with each other, to exchange information, so that, in the super-universe, in which the multi-verses were united, the total symmetry would be preserved—all of this is absurd). This speculation is very similar to Julian Barbour's *new theory of time.* I cannot come to terms with such speculations, because one can then assert anything, which is basically unscien-tific, because it is not verifiable, not *falsifiable,* and explains nothing. It shifts the expla-nation out into a world that is not accessible to us. For us (at least, for me), it is simply irrelevant whether there are other universes *somewhere* else. One can believe it or not, it may even be the case, but we cannot verify it, it changes nothing for us, whether there are multiverses or not. For me, with such speculation, one moves on to the same level as

---

[45] Stephen D. Landy et al., "The Two-Dimensional Power Spectrum of the Las Campanas Red-shift Survey: Detection of Excess Power on 100 h-1 Mpc Scales", Astrophysical Journal Letters, Vol. 456, p. L1–L4 xxx.lanl.gov/abs/astro-ph/9510146; Stephen A. Shectman et al., "The Las Campanas Redshift Survey", Astrophysical Journal, Vol. 470, p. 17–188, xxx.lanl.gov/abs/astro-ph/9604167

[46] Another article describing strong evidence for a high level of structuring of the universe can be seen here, in contrast to the often repeated claim "The universe is homogeneous on a large scale" (cf article in "nature" Jan 21, 1999: "The large-scale smoothness of the Universe" (https://www.nature.com/articles/16637). No, it is obviously criss-crossed by extremely fine structures. Astrono-mers also confirm in this study that the galaxies are connected to each other by very "fine" hot fila-ments, similar to dust threads on old attics (often misinterpreted as "spider webs"), see also Chap. 6 and these articles https://arxiv.org/abs/1709.10378v1 and https://arxiv.org/abs/1709.05024

religion. Something is asserted that can neither be proven nor disproven, one can only believe it or not. This is no longer physics, and it is not science. This is metaphysics. And it still does not explain what the *essence, the nature of time* is, which, in multiverses, has this direction *here*, and that direction *there*.

The assessment *not verifiable* should also be applied to Lee Smolin's assertion that time is *fundamental,* thus meaning that it exists independently of the universe, hence including *before* the beginning of our universe.[47] He speculates, like many other physicists, that there were many other universes *before* our universe, cosmoses in which nature could try out different types of physical laws. So, universes also experienced an evolution, Smolin says. They could do this because they reproduced themselves: From black holes, new *baby universes* would arise. At least Smolin is one of the physicists who holds *time* as something that actually exists, however, without proposing ways to help us understand what the *essence of time* is. It may be that, *someday* and *somewhere,* there were or still are other universes, but both the *someday* and the *somewhere* are totally outside our space-time, inaccessible to us. It is as if they did not exist. Speculations about the evolution of universes from baby universes help in no way further our consideration of what the the nature of *time* is. Because even if time is *fundamental* and exists *somewhere else* where other physical laws apply, we actually only want to know what the *nature of time is now,* here, with us. But it should be noted again that Smolin's view represents one of the positions in physics, according to which the time is *fundamental*. If that were the case, it would thus not arise from anything else or from other deeper reasons, but it would simply *always and forever* exist, even when there was not yet any entropy (and thus no arrow of time), also when there was only *nothing*, namely, absolutely nothing. No space, nothing. Only time? What then is the *essence of time,* which is allegedly always there, in the *Nothing*, regardless of whether our universe or other universes exist?

We have not yet been able to answer the question of the nature of time in spite of a lot of written material in articles and books by well-known physicists. We know what *mass* is, what *energy* is, and what a *distance* is. We can measure *time* (at least, we believe that we can, because we have clocks), but when we try to define the nature of time, we end up in a vicious circle: "Time is what we measure with the clock" (modified quote by Albert Einstein). This does not help us any further at all.

What Flows There When Time Flows?

So, let us start thinking again in a simpler way. We know what is flowing when we look at a river: water flows. But what flows when we feel and say that time flows? "Well, just *time* is flowing!" would be a usual answer. Then, my question again: What is that? Please

---

[47] Lee Smolin, In the *Universum der Zeit* Pantheon-Verlag 2015 (original issue 2013 „Time Reborn", Houghton Mifflin Harcourt, N. Y.).

do not take it amiss if I ask a few stupid questions now; I will write them in the exact form as I have been asking myself for years. Be patient.

We often say: "Time passes," but where or whither or how or when does it pass? Where does it come from? The arrow of time points to the future. Does it therefore flow from the past into the future? That's what we always say. What has time become then, if it has *already flowed* and passed by us in the present? What does it do in the future? Does it wait there for us to arrive in our present and then flow further into the future? Or, because we say: "Oh, how time flies!", does it fly into the future? How could it come from the past, then become noticeable and measurable to us in the present, if the past has gone and is limited? There have only been about 14 billion years of past so far; is there or was there a sufficiently large *reservoir of time* that can flow past us into the future? Because until the end of the universe, much more time will pass than has passed so far. I cannot imagine such a process, because the past is past. How the time can flow towards the future and through our present, meaning that, theoretically, it should come from the past, all that is beyond me.

Maybe the flow direction of time is the other way around? I'm just writing this as a thought experiment. Even if the *arrow of time* undoubtedly points to the future, the flow direction of time could be the opposite, just as the arrow of the river could be drawn from the mouth to the source, because we want to find the source, and our hike therefore goes uphill against the flow of the river. Our hike through life goes from the past towards the future: "Always follow the arrow of time," the direction of which, as we all know, is also uphill, with many difficulties. But the flow direction of time could then be the opposite of our hike, just like the river, time again flows past us in the present, but now, as a Gedankenexperiment, the flow of time is from the future into the past. Where did it come from then, if the future does not yet exist, but is only coming? Is there a vast amount of time waiting there (in the non-existent future) to flow past us in the present and then disappear into the past, as if into a sinking, to "pass away," as we formulate it?

Obviously, something is fundamentally wrong with our concept and understanding of time. Because if, logically, there can be no *time* in the future (which does not yet exist), and if it disappears into the past (*time passes*), then does it also not exist there, does it only exist in the present? Would the same be the case if time comes from the past (which, however, is already past!?), flows past us in the present and is passing away into the future? Would it then only exist in the present? How long, please, does the present last? I think everyone agrees that the real, physical present is extremely short, probably infinitesimally short, not even a moment long.[48] Is this perhaps an argument that time is

---

[48] H. D. Zeh formulates an interesting thought in his essay "Time in Nature" (German) (https:// www.thp.uni-koeln.de/gravitation/zeh/zeit-in-natur.pdf). He writes: "Similarly, the present (and thus also its apparent 'flow' towards the future defined by it) is apparently not a property of time itself, insofar as this can be understood as a concept characterizing nature. In contrast, neurobiology can show that processes separated by tenths of a second or even seconds are 'perceptual

an illusion? Or would both mental variants of the time flow rather be arguments that time is constantly being created?

If we stay with the analogy of a *river*, we should be more precise linguistically: It is not the river that flows, but the water. If we say: "The river is flowing particularly quickly today!", it is actually an imprecise formulation. The correct form would be: "The water is flowing particularly quickly today", and that "in the river bed". So, let's for a moment imagine *time* as a *river bed*. Does the river, the river bed, disappear when the water flows past us from the mountain down to the sea? No, the river, again, the river bed, remains; only the water flows past. And it does not disappear either, but ends up in the sea. As long as water flows in the river, the river remains. *River* is our term for what the water causes when it flows down in a stream from higher levels.

The phrase "time has passed" implies that it could disappear, but let me insist for a little longer that it no more disappears than the river, the river bed, in my example. If the river dries up, the river bed remains (as in my photo of the beach). And just as the traces of what flowed through the river bed are preserved, so can we at least partially *read* the past, as what flowed in time, in the past time, left traces in the present and also acts in the future. For the sake of simplicity, let's assume the following: *Time does not flow.*

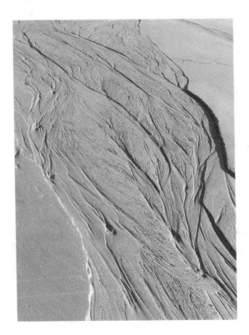

<hr />

packages' (that is, before their perception, a physically interpretable loss of information about the temporal ordering takes place). In this respect, the 'real' time continuum is just as fictitious as the (not violet-red) continuous spectrum of light. But it should be emphasized once again that this fictitious character does not mean secondary importance to something more fundamental or (apart from the purely subjective) better justifiable. " We must also be aware that the *perception* of time has nothing to do with the *essence* of time, more on this in the next chapter.

H. D. Zeh has also shown that time does not flow. He writes that the *flow of time* is a tautology. "The lack of a flow in the concept of time is unfortunately also presented by some physicists who should know better as a deficiency of this concept. Recognizing the logical error of such criticism does not require a scientist, because: 'Time goes, you say? Ah, no! Alas, time stays, we go.' (Austin Dobson) What we observe is only the 'simultaneity' understood as the correlation of momentary states of different motion processes." (cited by H. D. Zeh)[49]

What flows then, if we feel the flow of time, but it is not time itself that flows? What is the analogue to water, which flows in the *river bed*: what flows in *time*? If we can answer that, we know *the nature, the essence of time*.

The New Hypothesis About the Nature of Time

I now put forward the following hypothesis for discussion:

> Time is created by the flow of entropy. The production or export of entropy from open systems leads to an entropy flow through three-dimensional space, whereby the fourth dimension of spacetime, time, is formed. The time we measure is proportional to the entropy production, to the flow of entropy. Time is a kind of matrix in which entropy flows.

Previously, the second was defined as 1/60 of a minute or 1/3600 of an hour, which is 1/24 of a day.[50] The reference basis *1 day* already represents a connection with the entropy flow, because a certain amount of solar energy hits the earth on a day, driving a multitude of processes that produce entropy. Most of this is exported into space. So, a huge amount of entropy flows during a day.

Today, the second is defined differently, more precisely:

> "The second is defined by taking the fixed numerical value of the cesium frequency $\Delta \nu Cs$, the unperturbed ground-state hyperfine transition frequency of the cesium-133 atom, to be 9,192,631,770 when expressed in the unit Hz, which is equal to s^{-1}."[51]

The excitation and the radiation create entropy. The microwaves radiation emitted by the excited cesium-133 atom is associated with an increase in entropy, the frequency that we measure is the calibration of our time measurement. All kinds of clocks that we use are based on processes that produce entropy, including atomic clocks (see further below). Also, our sense of time, our *inner clock*, is the result of oscillations in a part of our brain,

---

[49] H. D. Zeh, "Time in Nature" (German) https://www.thp.uni-koeln.de/gravitation/zeh/zeit-in-natur.pdf.

[50] https://www.nist.gov/si-redefinition/definitions-si-base-units

[51] see previous footnote.

in some special cells there. This biochemical clock is also associated with entropy production, and thus entropy flow. More on this in the next chapter.

This means: *Changes* (in this case, the emission of radiation due to the excitation of the cesium atoms) *do not happen in time* or *during a certain period of time,* but the entropy flow associated with it creates the time, makes time come into being in the first place!

The essence of a river is the flowing of the water from the mountain down into the valley and then into the sea. The nature (the essence) of time is the flow of entropy, but not necessarily only from low to high values (as in the universe as a whole), but also from higher to lower values (as in non-equilibrium systems). So, it is not time that flows. Our common saying "time flies by" or "time is passing by" is inappropriate. Rather, it is entropy that flows. Both entropy decrease and entropy increase represent a flow that generates time. Since different amounts of entropy flow in different systems, *time is created* in each case. There is therefore no universal time, even if we represent it as such with our (atomic) clocks. Now we can also understand how the time *arrow* comes into existence and gets its direction: If the absolute value of entropy changes in a system, regardless of whether it increases or decreases, time is created, and we say that it progresses.

In my explanatory model, time is not fundamental, but, as one would precisely say, *emergent.*[52] This means that it arises through something; in my hypothesis, it is created by the flow of entropy. Even if it is *only* emergent (i.e., not *fundamental*), it is by no means therefore *not real* or even an *illusion.* There are a lot of emergent phenomena in physics, chemistry, biology and society that are undoubtedly very real, such as evolution, which was suddenly able to arise when inheritance became possible in the first living cells and organisms, and better or worse survival properties and reproductive capacity resulted from mutations. Emergent phenomena always arise on a higher aggregation level when elements of a lower level of organization of matter interact and create something new, like "surface tension", which I mentioned above as an emergent property of water, when enough water molecules are together so that a *liquid* is formed.

Perhaps the following example can illustrate the meaning of *emergence* to you even better: A single iron or copper atom has no electrical conductivity and is not a metal. After all, metallic conductivity is the transport of electrons on a scale of at least nanometers across many atoms. With only 1 single atom, no electron flow can occur, hence, 1 iron atom is not conductive. Even a dozen such atoms in a cluster do not yet behave like a metal. The very real phenomenon of *electrical conductivity* only arises in the case of gold when at least 55 gold atoms are very close together.[53] Other researchers came to the conclusion that at least 144 gold atoms would be needed before metallic properties would finally occur, 122 atoms do still behave just as a giant molecule or atom

---

[52] https://en.wikipedia.org/wiki/Emergence

[53] "When is gold a metal?" Physics Journal 1 (2002), No. 1, p. 20. (German)

cluster would, but not yet as a metal.[54] We don't need to worry about whether at least 55 or 144 gold atoms are required for the origination—that is, *emergence*—of the *metallic* property if we want to understand the concept that new phenomena arise on higher aggregation levels: When a minimum number of gold atoms are together and arranged so that they can exchange their electrons—that is, interact—, a new property arises that we call *metallic*. This cannot be derived from or explained with the properties of isolated gold atoms. And in the case of these material properties, it is again interesting to see that it is not until we investigate a metal with a particle size of at least 1 micrometer that the full metallic properties, as we normally know them, are present. Because, between the size of clusters of a few dozen or hundred atoms and below 1000 nanometers, they behave like *nanometals,* whose conductivity is limited by tunneling between the nano-particles.[55] On this new scale, on this even higher aggregation level, another new property arises (*emerges*). And all of these emergent properties are real.

Just as new phenomena (e.g., *conductivity*) emerge on a new aggregation level, requiring new terms (e.g., *metal*), we also use different methods to investigate them. *Time* is no different, being an emergent phenomenon. According to my hypothesis, it is a kind of matrix, in mathematical terms, the fourth dimension of Einsteinian spacetime. In this, entropy flows, comparable to a water flow in a riverbed, and the resulting "river" is also not just a reflection of flowing water, nor an illusion, but a real phenomenon caused by a larger amount of flowing water.

Time came into existence with the emergence of entropy. We can make this comprehensible by imagining how a river or riverbed comes into existence, namely, when enough water flows relatively coherently in a general direction from top to bottom. In this way, time comes into existence because entropy in the overall universe increases from low to high values, flows from low to high, or vice versa if we are dealing with non-equilibrium systems. Let us not forget: Even if, in open systems, such as a growing tree or an embryo developing into a baby, the entropy of this open system *tree* or *baby* decreases, it nevertheless increases in the surroundings and, ultimately, in the universe as a whole, because the entropy of these subsystems and of the entire earth is exported into the universe. Time exists because of entropy production, because of entropy flow in a three-dimensional space, in smaller or larger spaces. Time is a function of entropy flow, i.e., of the change in entropy in a certain volume.

---

[54] https://phys.org/news/2015-04-gold-atoms-metal.html,   https://www.sciencedaily.com/releases/ 2015/04/150410083516.htm

[55] P. Marquardt, Quantum-Size Affected Conductivity of Mesoscopic Metal Particles Physics Letters A 123, No. 7, 365–368 (1987), cf. also G. Nimtz, A. Enders, P. Marquardt, R. Pelster, B. Wessling, Size-limited conductivity in submicrometre metal particles. Similarities with conducting polymers? Synthetic Metals 45, 197 (1991). https://www.researchgate.net/publication/256143667_ Size-limited_conductivity_in_submicrometre_metal_particles_Similarities_with_conducting_polymers

There is no absolute time, which means that even if we were to define a second to be 9,192,631,770th times the period of radiation corresponding to the transition between the two hyperfine structure levels of the ground state of atoms of the nuclide ^{133}Cs, this would not mean that entropy production, and hence the time, in all subsystems of the universe is the same.

It is, *for our time,* irrelevant whether more entropy is produced in one system than in another, for example, in ourselves in comparison to an apple tree in the garden. It is irrelevant for us whether possibly less entropy flows in a glacier than in a growing snowflake, in a 5000-year-old pine tree growing very slowly in the Californian White Mountains less than in a fast-growing bamboo in China, so that the local time for the glacier or the pine tree would appear slower there. It is irrelevant whether—if the ice crystal in the glacier could experience entropy flow as *time*—time would appear to pass more slowly for it than for a snowflake, or for the cells in the ancient pine tree more slowly than for those in the bamboo. Globally or in the universe, the gigantic entropy production is relatively uniform, as can be seen in processes on the sun, in the movement of planets and in that of galaxies within galaxies and of galaxies relative to each other, as well as in the expansion of the universe. In a river, not every water molecule has the same speed, but the macroscopic flow speed of the river is measurable, and only changes with the amount of water, depending on rainfall.

The production of entropy, and thus the time as a consequence of the entropy flow, is not only not the same at different locations on Earth and in the universe, but is also not necessarily stable over time. Global and universal entropy production could become slower or faster, —we would not notice it, because we would still define the second as described above, calibrating the time by means of a frequency. For us, who live on Earth, who perceive nature here, live in it and explore it, a second is objectively a second. And what we call a day, is the time it takes the Earth to make one rotation. But that has nothing to do with the subjective feeling of time that sometimes makes us believe that time is racing by, and sometimes makes the seconds seem to stretch to minutes or hours; more about this in the next chapter.

The New Hypothesis is Verifiable and Falsifiable

Now, according to the theory of relativity, we basically have a way to test my hypothesis, because it predicts a slowing down of the flow of time at—in comparison to the *stationary* observer—higher speeds. However, it is not quite so simple, because it also predicts that time progresses more quickly in a weaker gravitational field.[56] Thus, two

---

[56] https://en.wikipedia.org/wiki/Time_dilation   and   http://www.astronomy.ohio-state.edu/~pogge/Ast162/Unit5/gps.html

contributions must be taken into account when we look at the time course in satellites in comparison to that on Earth. In GPS satellites, the atomic clocks run ahead by 38 microseconds a day according to the theory of relativity, an amount that would correspond to a greater entropy production than on Earth. These satellites orbit the Earth at about 20,000 kilometers altitude. Here, the influence of the weaker gravity (faster flow of time) prevails over that of the higher speed (slower flow of time). Anyway, space travel offers a way for my hypothesis to be falsified: In space, by comparing precise time measurement with the entropy flow of a suitable oscillating chemical reaction (e.g., a *chemical clock* or a *biological clock*), my hypothesis can be confirmed or refuted. Then, we will know more precisely whether it is appropriate. Although, in my opinion, the comparison of the atomic clocks on Earth with those in satellites is already a strong indication, because, in a trailing atomic clock, correspondingly less entropy was generated and exported in the form of heat radiation. But this probably does not convince the physicists who, according to the theory of relativity, expected the result in the form of advanced or delayed *time*. You do not necessarily see the connection with entropy at first glance, because physicists do not associate time *measurement* with *entropy production*. If you look closely, you will notice: All the clocks that have been designed so far generate entropy flow. We have to put energy into each clock, whether potential mechanical (with a spring or with weights), (electro-)chemical in batteries or, in the case of atomic clocks, electricity for the operation of the vacuum chamber, the atomic furnace and the microwave resonator.[57] These are all processes that generate a lot of entropy flow. If these atomic clocks run slower, less entropy was generated, i.e., in the vacuum chamber, the atomic furnace and the microwave resonator, and exported, if they run faster than the comparison clocks on Earth, more entropy has flowed.

Perhaps, however, the NASA twin study provides a more convincing indication: One of the two twins was in space for a whole year, the other remained on Earth. Fortunately, someone had the idea to *simply* compare what developed differently in the twins living in such different places and what did not. The motivation behind this study was, of course, not to study the nature of time. The motivation was to gain a deeper, more detailed knowledge of the health effects of staying in space. For this purpose, blood and excretions were examined in both twins before, during and after one of the brothers had spent time in space. Subsequently, one of the results surprised NASA very much; and, according to what I could read about it, it so far remains unexplained.[58] I'm glad that someone

---

[57] https://www.timeanddate.com/time/how-do-atomic-clocks-work.html

[58] https://www.nasa.gov/twins-study, complete publication in Science: https://science.sciencemag.org/content/364/6436/eaau8650, see short report in German: https://www.welt.de/wissenschaft/article161725255/Nasa-entdeckt-Jungbrunnen-Effekt-durch-Weltraumreisen.html (title translated: "NASA discovers fountain of youth effect by travelling in space"), see also https://en.wikipedia.org/wiki/ISS_year-long_mission, subsection NASA Twin Study.

at NASA came up with the idea of conducting these studies as well: The telomeres[59] of the space-traveling twin were longer than those of the twin who remained on Earth. This means, according to today's understanding of telomeres, that this twin aged more slowly than his terrestrial brother. According to my hypothesis, this would be expected: *Aging is accompanied by entropy production, lower entropy production would be consistent with slower aging, which could have been the case here.* This would be understandable in view of the time dilation in the space station. After all, it only flies at an altitude of about 400 kilometers, where the influence of the higher speed, which slows down time, exceeds that of the lower gravity. But there should only be a few milliseconds time delay in the course of a year.[60] Could this already have an effect on telomeres? After all, we could see a way in which my hypothesis could be verified, namely, in space, if certain cyclic chemical reactions, whose periods, total duration and entropy production are known, were carried out both on Earth and in space and then compared.

The Emergence of Entropy, and Thus of Time

Photons are timeless: they travel at the speed of light, do not change, and therefore have no entropy production. A photon *experiences* no time. Gravitational waves (which had been detected first in 2015) are also timeless; they therefore do not become *weaker over time.* This is also the case at the quantum level. For example, it is disputed whether or not protons decay over time. A half-life, if one exists at all, of at least 10^{36} years is assumed, but a proton decay has not been observed yet. This means that, even on the next higher level of aggregation above quarks, entropy does not yet exist, no entropy production takes place there either.

I assume that the phenomenon of entropy only (and *already*) arises at the level of atoms. Atoms can be converted into each other—nuclear fusion in stars, the production of helium and, ultimately, heavier and heavier elements, the decay of radioactive elements. On this level, we see becoming and passing away, that is, entropy production, and thus recognize the phenomenon of time.

---

[59] Telomeres are DNA sequences at the end of chromosomes. They consist of very long repetitions of nucleotides that have no effect on the synthesis of enzymes. Rather, they protect the chromosomes from being broken down, which would result in the loss of genes. They are considered an indicator of the aging process. See more about this here: https://en.wikipedia.org/wiki/Telomere, subsection Association with aging and more. There is a direct correlation between the initial telomere length in cultured cells and the number of cell divisions until the cells age. Patients suffering from progeria have abnormally short telomeres (https://www.sciencedirect.com/science/article/abs/pii/S0047637409000396). This and other findings suggest that the loss of telomerase function in somatic cells is related to the aging of a multicellular organism.

[60] https://de.qaz.wiki/wiki/Time_dilation#Relation_to_velocity

Quantum fluctuations, as they can occur hypothetically in the *nothingness* of the quantum vacuum, would be timeless. Before the beginning of the universe, before there were atoms, time did not exist, at least, not according to my hypothesis. Time is therefore not, as for Lee Smolin, *fundamental,* but *emergent.* It arises from something else, namely, from entropy, from the flow of entropy in space, and entropy only appears with the appearance of atoms. We thus realize that there are phenomena in our universe and around us on Earth that do not have time (photons, gravitation, quanta), while we live in time, can feel and measure it. Time is real. It arises from the interactions of substances in non-equilibrium systems and the entropy flow associated with it, it is emergent. There are many emergent phenomena in nature.

The laws of evolution and the phenomena of biology do not exist at the atomic level. A single iron (Fe) atom has no temperature. A single oxygen (O_2) molecule has no pressure. A single water molecule is not *liquid,* does not have a *water surface* and is not a gas, but simply a molecule. A SiO_2 molecule (a building block of silicates) is not *hard* or *solid.* A single stone, which consists essentially of silicates and is solid and *hard,* does not have statics: statics only arises and becomes relevant when we put many stones together to form a building or a bridge. One building does not yet represent a *cityscape,* it is not yet a city; only many buildings, streets and other elements together form this. We can therefore not reduce the properties of a city, a building or a bridge to the properties of SiO_2 and cannot develop a "unified theory of cityscape, statics and the SiO_2 molecule" (possibly even including quantum theory); such an attempt would be absurd and senseless.

Thermodynamics does not exist at the quantum level; only when a large number of atoms or molecules interact can we apply thermodynamics. Only at this level does the phenomenon of *entropy,* and thus *time,* arise. Time is not fundamental, but emergent, and it arises with the beginning of entropy production; not even at the Big Bang (if there was one) did it emerge, but only with the emergence of matter, which, according to common ideas, did not happen "at" the Big Bang, but only "later"—a strange statement, because time did not yet exist. But if we want to imagine the origin of the universe as described by Gunzig, Géhéniau and Prigogine (see Chap. 6), then this was the case at the *moment* when matter and space came into being.

This is how entropy production and what we call *time,* something that we can feel, sense, and measure as such, began. Time does not pass, neither does it flow. I think this is a linguistic confusion. The river does not pass while the water flows (only if it does not flow anymore, then no river is there, but a dry river bed remains). The formulation "flow of time" creates a misleading image: what is it that is supposed to be flowing there? It obfuscates the relatively simple qualitative knowledge: time is, greatly simplified and transferred to an everyday image, something similar to a riverbed, but in which entropy flows instead of water. Entropy flows in a matrix that represents the fourth dimension in the four-dimensional spacetime of relativity theory.[61]

---

[61] I cannot formulate this knowledge in quantitative form (yet?).

Now, we could also try to resolve the contradiction between the missing arrow of time in relativity theory and the reality of time, as well as the clear direction of time: On the level of photons (light speed) and gravitational waves (also light speed), there is no time, thermodynamics is not applicable, light and gravity *experience* no entropy production. Nevertheless, we can perceive light and gravity, and they act upon us and our environment. The timeless gravity and light—as described by relativity theory—act on the higher aggregation levels of matter. That is where entropy flow takes place, and thus time. The so-called *paradox* between relativity theory and non-equilibrium thermodynamics with its arrow of time does not exist. The theory of relativity describes space and time from the perspective of observers who move relative to each other, as well as the influence of gravity on space and time. It does not describe the becoming and passing of matter, nor does it describe the *origin* of matter, and thus entropy and time as a consequence. For relativity theory, space and time simply exist already. It is not a *theory of everything,* any more than the thermodynamical theory of non-equilibrium is. We need both (and a few more, like quantum physics, cosmology and the theory of evolution) to understand the world in which we live.

The End of Time

We have considered the beginning of time at the beginning of the universe and my hypothesis of the origin of time through the flow of entropy. Thus, it is now *time* to also think about the end of time. At the end of the last and at the beginning of this chapter, I wrote of *black holes in which time stands still.* It is probably a widely held view in science that the majority of the entropy of the universe is to be found in black holes, as the mathematician and theoretical physicist Roger Penrose, who was honoured with the Nobel Prize in 2020 for his work on black holes, wrote.[62] In their most important publication on this topic, Hawking and Penrose prove the possibility of the existence of these giants and the end of time in the singularity of black holes.[63] According to their theory, space-time ultimately only collapses in the centre of the black hole, so that time no longer exists there. In the light of my considerations, one would also have to think about at what depth in these voracious giants of the universe changes no longer take place, and therefore also no entropy is produced. There, according to my model, time would end. This could already be the case within the event horizon. Investigating this would be a worthwhile but not easy task for theoretical physicists and non-equilibrium thermody-

---

[62] Roger Penrose, „Zyklen der Zeit", Springer Spektrum 2011, Softcover 2013, p. 79 and 147. Original english edition „Cycles of Time—an extraordinary new view of the Universe", The Bodley Head 2010.

[63] S. Hawking, R. Penrose, The singularities of gravitational collapse and cosmology. In: Proceedings of the Royal Society. A. Bd. 314, 1970.

namicists. But one thing is already clear: In equilibrium, a system no longer changes, there are no movements, no material changes. In equilibrium, time stands still.

This is, of course, only the case for equilibrium systems as long as they are *closed*. There may be no external influences whatsoever, neither in the form of energy nor of matter. If this should happen at some point, something changes in the system, and it reaches a new equilibrium level. This applies to a sugary water solution in a closed can just as much as to a black hole far out in the universe. As long as nothing from the outside acts on the system, there is no entropy flow, neither in nor out. Time stands still. If, for example, something flows in, such as a neutron star into a black hole, entropy flows in, and a beginning and an end of time have occured there again.

If my hypothesis about the nature of time should prove to be helpful for understanding time even after further experimental testing, as may be indicated by the NASA twin study, I have fulfilled my promise from the beginning of this chapter: We have gained knowledge without having to pay for it with a loss of clarity. Now, we just need to understand a little better how we humans experience, feel and perceive time. If you would also like to think about this with me, please turn to the next chapter.

Our Perception of Time

8

Abstract

It is important to understand that our perception of time has nothing to do with what the nature of time is, just as our perception of color has nothing to do with the nature of light or hearing a sound with the nature of sound. We learn how the body organizes rhythms, what *clocks* the cells have and how and where the sense of time is generated in the brain. Finally, it is explained why, for many people in old age, time appears to *pass* more quickly. A close connection to the nature of time (see Chap. 7) becomes clear.

Our intensive occupation with the causes of the emergence of chance and with the nature of time would be incomplete if we did not also turn to the time in ourselves. After we have seen that chance, coincidence and time have entropy in non-equilibrium systems in common, now, only time should interest us, no longer chance or coincidence, nor any accidents. There are 2 topics to which we will turn in this chapter,

- the control of the various rhythms in the cells and organs of the body and their sensory and neurological processing, as well as
- the psychological aspects of time perception.

Of course, I am not asking you that you become an expert in chronobiology[1] or chronopsychology[2] (if you are not already), nor am I an expert in these science fields, but together, we can now gain a small insight into them.

---

[1] https://www.spektrum.de/lexikon/biologie/chronobiologie/13999 *and also* https://de.wikipedia.org/wiki/Chronobiologie.

[2] https://www.spektrum.de/lexikon/biologie/chronopsychologie/14010 *and also* https://de.wikipedia.org/wiki/Chronopsychologie.

B. Wessling, *What a Coincidence!*, https://doi.org/10.1007/978-3-658-40671-4_8

Let me point out beforehand that what the body does with and in time is quite different from how we perceive it. The phenomenon of time *sensation* must also not be confused with the *essence of time,* just as the sensation of color must not be confused with the essence of light and its frequency spectrum that hits the eyes, nor should our perception of sounds be taken for the nature of sound, i. e., the vibrations of the air that arrive in our ears. Light waves and air vibrations are only processed in the brain into sensations and perception that we call *color* and *sound* via suitable sensory cells and neurological connections. The cognitive researcher Fred Mast writes, among other things, in his fascinating book about "imagination" (translated by the author): "One must be aware that the physical stimulus *causing* it [a sensation of color or sound] is not known to us [the author probably means: "not conscious"; note by the author]. [...] We have to decipher the physical stimulus. [...] Assumptions [of the brain; note by the author] about the stimuli play an important role in the perception process." [3] We therefore have to distinguish what the body receives in terms of external (or internal) physical stimuli about time, how the body controls times and rhythms, and, on the other hand, how the brain perceives time and makes it experienceable for us, and why we rarely feel time as progressing at a constant speed.

Let's start with what our body does over time and in time. We all feel our inner rhythms, our bio rhythm, every day: We feel particularly vigorous in the morning (if we have slept well). At noon, we are not as productive and would rather take a break (the southern Europeans take a siesta). And whether we are the so-called *larks* or the *owls*, we all have a relatively stable sleep/wakefulness rhythm, although different ones. We go to the toilet at about the same time every day to "drop a packet". And anyone who has flown to America or Asia knows jet lag. We consider it quite normal—but how does my body still know, when I have landed in China, what time it is in Germany? Why does he still feel like he's at home in the middle of the afternoon when it's dark here in China and I should go to bed and sleep so that I'm fully rested the next day? Whatever the reason, I can not, although dog tired, fall asleep until three o'clock in the morning or even later. Three o'clock in the morning in China is only 8 pm in Germany (if we have not switched to summer time). My body claims it is only 8 pm, and still will not let me fall asleep. Only after a few days does my bio rhythm also arrive in China, with my bowel movement even taking a few more days after that.

---

[3] Cf. Fred Mast, "Black Mamba oder die Macht der Imagination", Herder 2020, pp. 53 and 54 (German).

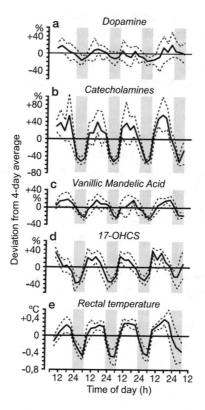

Many body functions take place in a wave-like manner. The stomach has a 45-s rhythm. The heart beats about 60 to 80 times a minute. The concentration of various substances in urine is shown in the graph above; these waves have a period of about 24 h.

The rhythm of the excretion of various substances in urine can also be seen in the graph above. Black line: Measurements averaged over six test subjects; the standard deviations between the values of the individual test subjects are shown dotted. The values were recorded every three hours over four days. *a–d* Profiles of dopamine, catecholamines, their breakdown product vanillin mandelic acid and of 17-hydroxycorticosteroids (17-OHCS). *e* The diurnal variation of rectal temperature. The change between day (16 h, white) and night (8 h, gray) is indicated. (From Wisser et al. 1973, p. 244; with kind permission from © Springer Nature 1973. All Rights Reserved).[4]

---

[4] See the following footnote, page 3, Fig. 1.1.

For all of this, there is something in every body cell that functions like a *clock,* these are oscillating biochemical reactions that are controlled by the genes.[5,6] The so-called *circadian rhythm* is responsible for practically everything in the biorhythm. This is a rhythm that spans "about one day" ("circadian"). The discovery of the *inner clock* was honored with the Nobel Prize in 2017, while I was working on a first draft of this chapter.[7,8]

This biological clock is based on complicated material feedback loops in a part of our brain, the *suprachiasmatic nucleus,* SCN.[9] That's where our biorhythm is generated, calibrated from the outside by daylight and other signals, and which coordinates all other *clocks* in the cells or even controls them. So, the SCN is responsible for the fact that, after arriving in China, I feel jet -lagged for a few days.

The underlying biochemical processes are extremely complicated and by no means fully elucidated in detail. However, the principle is now largely known. Simply put: A certain gene causes the synthesis of a protein that stimulates the production of another protein, which, in turn, inhibits the production of the first. A feedback sequence takes about 24 h, which is extended to about 25 h when the light-dark rhythm of the day-night cycle is artificially switched off. Protein synthesis and degradation—and all the rhythmic processes controlled by them in cell organelles and body organs, including those of cortisol and melatonin—are chemical transformation processes, i.e., associated with entropy production and flow. The melatonin concentration in the body is high during the night, while that of cortisol is low, the opposite is the case during the day (i. e., during the day, the melatonin level is low, the cortisol level high): A constant substance synthesis and degradation, which must first be re-adjusted—by means of daylight and eating habits. That's why I can't sleep well at night during the first few days in China.

To adjust more quickly, it is advantageous to arrive in the new time zone in the morning, to spend time outdoors during the day, and also the next day if possible, preferably with walks or longer leisurely hikes. With lots of daylight, the SCN is recalibrated, and the slight physical activity also helps. Then, the switchover goes faster. That's how I do it anyway. I consistently avoid taking business appointments immediately after arrival in

---

[5] The book by G. Eichele and H. Oster "Auf der Suche nach der biologischen Zeit", Springer 2020, (https://link.springer.com/book/10.1007/978-3-662-61544-7) provides an excellent detailed overview (German). A compact overview by Eichele et al can be found here: https://onlinelibrary.wiley.com/doi/10.1002/bies.201500026. An easy-to-read introduction in English can be found here: https://www.intechopen.com/chapters/68556

[6] https://en.wikipedia.org/wiki/Circadian_rhythm#In_mammals.

[7] https://www.nobelprize.org/prizes/medicine/2017/press-release/

[8] https://www.scientificamerican.com/podcast/episode/nobel-prize-explainer-circadian-rhythms-oscillatory-control-mechanism/

[9] G. Eichele, H. Oster, loc. cit. p. 111., cf also https://biologydictionary.net/suprachiasmatic-nucleus/

Asia or the USA. I take two days to adjust and spend as much of these days as possible in nature, even if it is, for example, in Tokyo, in large city parks, where local birdwatchers also gather at dawn.

All the countless biochemical clocks in the body cells are thus also linked to entropy flow. This flow therefore generates time in the body, too, in each cell. The *inner time* is then the matrix in which entropy flows. This was synchronized with the rotation of the earth, i.e., with a day's length, during the course of evolution and is felt by us as *a day*. We are mostly not conscious of this inner clock, but we feel the different rhythms during the course of a day.

How is our *inner clock* converted into a sensation of time? Where and how exactly the sense of time is localized or the course of time is perceived and converted into a feeling has not yet been thoroughly researched. The research activities dedicated to this question so far have led to numerous different models.[10] It is a well and broadly accepted idea that these are bodily processes that cause the perception of time in the brain. These bodily processes send signals to the brain—so that we can perceive heat, pain, hunger, thirst or a smell, which then associates such signals with a *feeling*, in this case a feeling for a period of time. As we know from many sensations (such as *pain,smell,hunger* or *cold*), the translation of signals from the body into sensations in the brain is by no means error-free, often being inaccurate or downright misleading. For some states in the body—such as blood pressure—we don't even have sensors that could send a signal to the brain. But for time sensation, they obviously exist.

According to a research group in the Department of System Neuroscience, Division of Cerebral Integration, National Institute for Physiological Sciences (NIPS), Japan, it is probably the *insula* in the brain where the sense of time is generated.[11] In this part of the brain, all physiological conditions of the body are processed. Here, the sensations are generated that inform our consciousness about the state of the body. Magnetic resonance imaging during tasks with a time reference—for example, reproducing sounds of different lengths—shows activity there. Based on such and other studies, the hypothesis was put forward that the accumulation of changes in the states in the body causes the sensation of duration. One such constantly recurring change is the heartbeat, which was analogously influenced by temporally structured tasks during the studies!

Scott M. Hitchcock particularly emphasizes the fact that we accumulate recollections in our memory, that is, information, and also arrange them in time.[12] We all know that

---

[10] M. Wittmann, The inner Sense of Time: How the brain creates a representation of duration, Nature Rev Neurosci 14 (3), 217–233, see also https://www.nature.com/articles/nrn3452, siehe auch: https://www.researchgate.net/publication/283617449_The_inner_sense_of_time_How_the_brain_creates_a_representation_of_duration.

[11] https://www.sciencedirect.com/science/article/pii/S0028393222000045?via%3Dihub

[12] Scott M. Hitchcock, "T-computers and the Origin of Time in the Brain", NeuroQuantology 2003; 4:393–403, cf. https://www.jneuroquantology.com/index.php/journal/article/view/26/26

this is not necessarily a reliable arrangement. He emphasizes that a memory of something is always based on a *change* ("'Change' creates signals that carry information."). This is, of course, correct, because an event that one remembers is a change compared to what was before or what was not. If nothing changes, there is nothing that the brain can store new memories about.

Did you notice? I quoted the term *changes* from the article by Marc Wittmann! (See Footnote 10) Hitchcock formulates it similarly, but also refers to the term *information*. You will surely remember that changes are associated with entropy flow. The same is true when accumulating information, because an increase in information is known to be identical to the formation of structure, that is, a decrease in entropy through entropy export. And thus, with the entropy flow, we have already arrived again at my hypothesis about the nature of time.

In many MRI studies on mammals and birds, it has been observed that "climbing neural activation", as referred to by chronobiologists, takes place in the insula during tasks with temporal or rhythmic structure. It is assumed that body signals serve as an internal reference for estimating the duration of external events. As already mentioned, however, the resulting sensations are anything but objective, just as sensations about other states in our body are not.

We all experience phases in which time seems to pass too quickly (vacation), and other phases in which it seems to progress too slowly. As children, we experienced this while waiting for Christmas, and then later, as teenagers or adults having freshly fallen in love, waiting for the objects of our affection to show up at our designated meeting place. In addition, many people we know tell us that time seems to pass more quickly as we age. Apparently, this is actually the case: In a study, it turned out that, the older the respondents were, the more quickly "the last ten years", in particular, seemed to have passed.[13] "In the ever-constant daily grind between job routine and family routine, only few events stand out that we can remember in retrospect," to quote Marc Wittmann.[14] "But the more that memories, especially beautiful ones, also rise up, the more fulfilled and longer a period of time seems in retrospect." The paradox, as he says, is: The more interesting and beautiful the experiences in a phase (e.g., of vacation) are, the faster time seems to pass while we experience it, but, in retrospect, it seems to have passed more slowly. I can't confirm that for myself, at least, not as a general rule. Of course it happens to me that I am involved in an interesting conversation and suddenly I notice: "Oh, it's already 11 o'clock! I would not have thought that!" If I experience something new, if I am very attentive not to miss anything, if I work very intensively on something new, I do not particularly feel time passing quickly during these phases, and in the memory of

---

[13] https://www.psychologytoday.com/intl/blog/sense-time/201604/the-passage-time-across-the-life-span.

[14] https://www.welt.de/gesundheit/psychologie/article123344063/Warum-die-Zeit-mal-rast-und-mal-nicht-vergeht.html

these phases, time also seems to me to have passed slowly. This may be very different for different people. And, of course, I also have such phases from time to time in which I am busy with a lot of routine activities and time seems to pass quickly for me, and even more quickly in retrospect.

In my experience, stepping out of one's comfort zone to do and experience new, unfamiliar things has another interesting side effect: One lives more intensely and consciously. I wrote about this in my diary in July 2004:[15]

> An insight into the sense of time
> Everyone knows that, as one gets older, time seems to go by faster, the days, weeks, and years just fly by. Surprisingly, this apparently does not apply to intensely experienced periods of time: The days I spent with the whooping cranes in Aransas, Hokkaido, or recently the weekend spent in Brandenburg—and now, surprisingly, the vacation. Contrary to my expectation, the three weeks afterwards did not fly by in a flash, now in retrospect, as I write this, they seem even longer.
>
> I realize that, the older one gets, the more routine tasks one performs—breakfast, driving, working in my company—so much is known and routine, that much of it is done relatively unconsciously. But (for me) new things, like diving, stone statues (*moai*) on Easter Island, black pearls, cranes, my first sighting of a greater spotted eagle, or other paradisiacal hours or minutes, they are experienced more consciously, because everything requires or stimulates attention. As with a child, for whom everything is new, who must and can experience everything very attentively.

Many times in my life, I have left my comfort zone, which is, for me, just as comfortable as it is for all people, and it is always difficult for me to break out of it. So it was with the crane research mentioned above in the diary in Texas, where I roamed all alone in a 500 km^2 large wild reserve in the darkest hours of the early morning, freezing terribly at night before going to sleep and then again after getting up.[16] I am really not an outdoor person like Rüdiger Nehberg, even though I like being outdoors a lot, but alone in the wilderness? And sleeping outdoors at practically minus 5°C? That required a big step out of the comfort zone, and it was great. My jump into China, which was required for my business, was even harder for me, I was terribly afraid of it. But now, I remember this long phase of life as having passed quite slowly, while, at the same time,

---

[15] From 1996 to 2006, I kept a diary during the 10 years it took to build my next company (Ormecon), years that were extremely busy, financially risky, and demanding, with enormous additional progress in research and marketing my new "Organic Metal" technology. Every few days or weeks, I would write down my thoughts on the progress or the constant setbacks, and I dealt with my stress in this way.

[16] cf. my book "The Call of the Cranes", Goldmann 2020 (https://www.bernhard-wessling.com/the-call-of-the-cranes-short-intro).

being enormously valuable and instructive.[17] These and many other breakouts from my comfort zone have enriched my life and ensured that I have never so far needed to complain about *time passing too quickly*. These were kayak trips alone and with my sons at sea, or an eight-day tour with my older son through a really deserted national park in northern Europe, where, after four days, we were about fifty kilometers from the nearest road and the nearest house. Several-day stays in a wooden cabin in northern Iceland in autumn and winter, with long horseback rides alone in the wide wild landscape, as well as actively participating several times in the sheep drive-off in autumn, again by horse, in the mountains and river valleys. Not least of these examples was the mentally very difficult task of overcoming of my fear, no: my panic, when diving, which then helped me to experience the beauty of our underwater world—unforgettable encounters, numbering by now about six hundred dives, mostly in areas of our world oceans that had seen little to no diving before.

By chance, I later found an article that, in a mathematical way, sets out a similar conclusion to the one I had noted down for myself in my diary: In old age, time seems to progress faster because the number of really new and memorable experiences is lower than in younger years and constantly decreasing.[18] A month during childhood is perceived as longer than a month in old age. Because of the lower density of new experiences, the authors of the article in the footnote speak of a perceived "dilution of time." However, the authors do not claim the inevitability of such a development with their mathematical model. Once significant new events occur, regardless of whether one is young or old, the potential for subsequent new events is suddenly much greater and the sense of time "rejuvenates". I can fully confirm this for myself, because my life in China began when I was in my mid-fifties, my professional/business reorientation while still in China after selling my company in my late fifties, and my return to Germany, where everything started anew here, in my late sixties, as well as my move shortly afterwards into an older house that I had bought, accompanied by many significant personal changes. None of this inspired a feeling of "oh, how time flies by, always faster!"

The *job routine* mentioned by Wittmann in the interview quoted above is replaced, for most people in old age, by another daily routine with few new impressions. If you want to break out of this dilemma, the consequence is clear: Avoid falling into routine as often as possible; change your environment, processes, tasks, goals, procedures; the more new and different ways of living I create, the more intensively I can experience time, my life.

---

[17] cf. my book "The Jump into the Cold Water", scheduled to be issued by Eulenspiegel-Verlagsgruppe in 2023 (German).

[18] F. T. Bruss, L. Rüschendorf, "On the Perception of Time", https://www.researchgate.net/publication/40850494_On_the_Perception_of_Time, Gerontology, 56 (4), 361–370 (2010), https://www.researchgate.net/publication/40850494_On_the_Perception_of_Time. cf also https://lifehacker.com/why-time-feels-like-it-s-flying-by-and-how-to-slow-it-1745852093

Chronopsychologist Marc Wittmann researches the perception of time. During an interview with WELT, he, in his further capacity as a *waiting consultant* (i.e., someone who advises people who lack patience, feel bored when waiting, and cannot cope with it), he spoke about change in the perception of time that people experienced during the first months of the Corona pandemic:[19] "In spring [referring to the year 2020; note by the author], many people experienced that time passed incredibly quickly. Because every day is the same. People are at home in the home office, and Monday and Sunday are no longer different. Sunday used to be the day when you went for a bike ride or did something with the family. Monday was then a stressful work day. Or you had fixed appointments, dates. You went to a soccer game or something. All of that fell away. Every day was the same. Because of this monotony, because nothing special is stored in memory anymore, time seems to have passed quickly when looking back. If you experience a lot of variety over the days, your subjective time stretches out in retrospect. ... Those who felt uplifted in a good social situation—because they lived together with their family or with friends in a shared flat and were satisfied with their environment—for them, time passed by in an almost accelerated manner. Those who were very dissatisfied and lacked involvement, they felt a lot of boredom. In this case, as well as for depressives or people with anxiety disorders, time passes more slowly. "

To me, this time since the beginning of the pandemic has seemed to progress very slowly and intensively, every week, every month, even though I don't suffer from depression, anxiety disorders or boredom. As I work on suggestions for changes in this chapter from the lector (February 2023), I look back to March 2020, when my book on crane research, which I have been pursuing part-time for decades, was published.[20] On the very day that it was published, everything closed, bookstores included, and in the media, the Corona pandemic was by far the No. 1 topic, so there were hardly any reviews, and all scheduled book presentations were cancelled. For me, it was a very intense year, similar to 2021: I constantly had to re-organise, reschedule, come up with new ideas about how to make the book known, which, in my opinion, it deserved to be—because it is about a new image of the intelligence of cranes and of birds in general, about their flexibility. I also plead for more respect for nature, for all animals, for a greater understanding of the urgent need for species conservation, the preservation of biodiversity.

In the above-mentioned interview, Wittmann further states that, in our modern times, information flows more rapidly, travel has accelerated, work processes have become faster, all of which leads to a feeling that time is passing more quickly. "City dwellers

---

[19] https://www.welt.de/wissenschaft/plus220730196/Psychologie-Was-Ihr-Zeitempfinden-ueber-Sie-verraet.html WELT online from 27. 11. 2020, accessed via "WELT plus" subscription (German).

[20] B. Wessling, Der Ruf der Kraniche (German edition Goldmann 2020), see also http://www. bernhard-wessling.com/the-call-of-the-cranes-short-info, English edition "The Call of the Cranes", (Springer 2022), see also https://link.springer.com/book/10.1007/978-3-030-98283-6.

live faster than rural dwellers", and they are more impatient than those in the country-side.

In my opinion, you can't state such a thing so categorically. The perception of time is very individualized, from person to person, from situation to situation, no different than the perception

- of a distance (what is a stone's throw for one person is too far for someone else, even though both are about the same age and both are in good physical condition).
- of a weight ("it's so heavy!"—"no, quite light").
- of a loud sound ("unbearable!"—"wonderful, this Beethoven symphony, it has to be played so loud!").
- of a smell (pleasant, unpleasant; delicate or overpowering?).
- of a taste (wonderfully fresh, unpleasant, interesting, uninteresting).

When we dip a hand or a foot or throw our whole body into a river, we feel how the water flows down the mountain. We have discussed in this chapter how we feel time, the flow of entropy that creates time. I have sketched the state of research that ultimately leads to the knowledge that entropy is produced in the body, in our organs and in the cells. In addition to the *biochemical clocks* in the cells and in the brain (which are largely not conscious), there are other *clocks:* Our heartbeat, our breathing are rhythmic processes that, on the one hand, generate entropy and, on the other hand, can become conscious. Hearts and lungs pulsate quite regularly. When we walk—right, left, right, left, 1, 2, 1, 2 …—we feel another *clock* ticking. These types of clocks are not exact, but they give us the feeling of time. When we wake up, we are hungry and have breakfast; after a certain *time,* which we could experience because of our heartbeats, or because we went shopping, we are hungry again and prepare lunch. We cut cucumbers, back and forth, back and forth—you can get many slices from a long cucumber (1, 2, 3, 4, 5, 6, 7 …), representing a time sequence; we peel potatoes, fill a pot with water and much more—all processes that take place *in time,* processes that cause entropy flow. In the evening, it gets dark, we experience and feel the day/night rhythm. We observe tides, seasons, the growth of trees, the blooming and wilting of flowers. So, we feel entropy flow as a progression of time. All these processes generate entropy. With our birth —actually already at conception —we were thrown into the time-matrix of space-time. Being and passing is entropy flux, and it flows in the "river bed" of time.

One further method for preventing feeling that time is passing so quickly in old age is to break out of the routine that has grown during one's life. You leave your comfort zone and do something unusual. My 13 years in China forced me every day to break out of the routine. In the last years of my life in China, we undertook several major trips to areas that are very little developed for tourism. These were not backpacker tours, but we were very far outside our comfort zone. Here is a view from a hill in the Gobi desert. These were unforgettable weeks full of unique impressions that have noticeably "rejuvenated" my feeling of time. (Photo of the author 2016).

Final Remarks

We have undertaken a long mental journey in this book and have now arrived at our destination. I was able to deliver to some of you, I hope, a new or, at least, a deeper understanding of the world we live in. I may have shown some readers some new aspects. Some will have been surprised or even disturbed that our world is not *in equilibrium* at all. We don't have to worry about it, it is nothing new, it has always been like this; it's just that we humans tend to suppress it. This does not change the fact that those who do not know, as well as those who do know, are always confronted with chance and coincidental events. The non-linearity of all the different non-equilibrium processes and interactions is simply unavoidable and unpredictable. Therefore, coincidental events are quite normal. Catastrophes are terrible, many accidental events are existentially dangerous or even deadly. Sai Weng taught me that, as long as I can overcome a "fateful" blow to some extent, something good may develop from it. And he taught me to take lucky coincidences not only with humility, but also with caution: They can also have a negative effect.

The entropy flow in and out of open systems has turned out to be a key element for the causes of chance and coincidence and for the nature, for the essence, of time. Only in non-equilibrium systems far from equilibrium is something in motion, and this is not linearly the case. Entropy flows, structures are built, we can observe changes all the time, and entropy is exported. This results in situations with the possibility of forks in the road (bifurcations), and chance and coincidence surprise us again. Furthermore, the flow of entropy creates time. Without non-equilibrium, there would be no world, and no us either. We have thus also received chance, coincidence and time as a gift. In equilibrium systems, nothing happens any more once equilibrium is reached: Entropy has reached its maximum value, time stands still. This is the case in a black hole, just as it is in a watery sugar solution that we pour into a glass, sealing it so that the water does not evaporate – but that would also be just another way into another equilibrium in which only the sugar crystals would remain, so that time would also stand still in them, if it were not for some bacteria enjoying the sugar.

A critical reader[1] of one of the previous versions of my manuscript wrote to me during the course of a longer discussion: "But yes, the world is in approximate equilibrium, just not in *thermodynamic* equilibrium (according to Prigogine)." I asked him what kind of equilibrium then? His answer was an example: "If you go to bed at night, you do so with the certainty that you will wake up in the same bed in the same room in the morning. So, the world is approximately in equilibrium during your night's sleep." I replied to him by email: "I completely disagree, because I myself am already a non-equilibrium system (including in the sense of Bertalanffy!): I have eaten and drunk during the day (so I have taken in energy), while I continue to digest during the night, so that I excrete waste products in the morning (and thus a lot of entropy). And even during my sleep, I export a lot of entropy by radiating body heat. So, I am a typical open system in the sense of Bertalanffy and Prigogine. But the world around me was also not in equilibrium during the night, not even approximately: The temperature had dropped, in the morning, it rises again, the weather has changed, ebb and flow have taken place (definitely no equilibrium processes!), etc." I had not even considered the outbreak of a war like the Russian attack on Ukraine, which happened many months after the email exchange cited above.[2]

So, this critical reader and I had the problem that I had already discussed in Chap. 4 under the keyword *dynamic equilibrium*: We have different ideas of what *equilibrium* means. I hope I was able to convince you that it is necessary to use clear and unambiguous terms. We can not, if we want to understand our world better, use the same word *equilibrium* for two completely opposite phenomena or states. Equilibrium and non-

---

[1]A mathematician, and theoretical physicist, and emeritus university professor, who has dealt a lot with chance, coincidence and time; due to his critical remarks – for which I am very grateful – I had to carry out quite a bit more further thinking and research, the results of which are reflected in the book.

[2]When we discussed this, we were still a year and a half away from 24.2.2022. When I woke up that morning, Russia had attacked Ukraine on the orders of its President, Vladimir Putin. That showed that the security system that was set up in Europe after the Second World. War was "more wobbly and rotten than the West wanted to admit" (as the German weekly magazine DER SPIEGEL wrote on 25. 2., see https://www.spiegel.de/ausland/wladimir-putin-und-sein-krieg-in-der-ukraine-der-ang-riff-der-die-welt-veraendert-a-7720502b-38d1-4963-a331-3363242cec18). It seems as if the world has fallen back into a phase of ruthless superpower politics, with a very non-linear step: An attack war against a sovereign state, a brutal violation of the Charter of the United Nations, a violation of international law. I would like to refer once again to Chap. 6, subsection *The Birth of Chance and Coincidence in Non-Equilibrium*. There, I explained that not every event, not every non-linearity is a chance or a coincidence. Not even this war of aggression. But out of a chain of countless coincidences, presumably occurring many years before, this war of aggression emerged in an apparently inevitable way (transition from chance to necessity). Inevitable because Russia's President Putin could not be dissuaded from his decision, a decision that can not be justified by anything. *Necessity* only from the view of the Russian government and its loyalists. For the people in Ukraine, for the country, for peace in Europe and the world, for the fight against hunger, a disaster, a catastrophe.

equilibrium are simply opposites, like minus and plus, like end and beginning. So, we should refer to them with different terms – so, the world becomes clearer.

Time arises, according to my hypothesis, in the non-equilibrium due to the flow of entropy. The thought came to me more than twenty-five years ago. At the time, I did not pursue this thought any further, partly because of lack of *time*, partly because I thought other people, especially non-equilibrium thermodynamicists from the Prigogine school, would have already come to similar thoughts. But that was obviously not the case. I know that I am taking a risk with the publication of my hypothesis about time; the same already with my explanations about the connection between chance, coincidence and non-equilibrium, which some scientists may not agree with. I am happy to take this risk.

Certainly, such readers who have a different idea of the nature of time will find my new hypothesis – if they react neutrally-friendly – strange. I look forward to *constructive* discussions. Let's wait and see if there are any other indirect hints or even direct experimental evidence for my proposal in the coming years, or for the refutation of my thoughts. Both will bring progress in knowledge.

Whatever the outcome of this very exciting question – the deep engagement with chance and coincidence and with the nature of time has enabled me, and hopefully also you, to gain many insights into the properties and behavior of non-equilibrium systems, and thus into our lives and our environment.

This understanding can help us to better understand and experience the world in which we live. This affects the world on the micro, the nano and the molecular scale, in cells and organisms. On a large scale, such as in ecosystems and our human societies. And on the largest scale, in the universe. And at the same time, how strange: The better we understand it, the more we realize how infinitely complex and mysterious our world is. The more questions we can answer, the more new questions open up before us. How beautiful and ever new it is again and again!

A few days before the start of my proofreading of the German edition of this book, I heard on the radio (of course, by chance) the following poem by Marie von Ebner-Eschenbach. In my opinion, it very well describes the complexity and, at the same time, the beauty, but also the reality of becoming and passing away in our world in a poetic form – similar to my photo of the Baltic Sea beach on the next but one page. This German poem was translated by Timothy Andès[3] (in a productive poetic exchange with the author Bernhard Wessling so that the poem's message could become optimally reflected and still sound like a poem).

---

[3]The German poem was translated here by Timothy Andès, a "rhyming translator-poet": https://www.timothyades.com/. He has translated dozens (if not hundreds) of poems from German, French and Spanish into English. B. Wessling reviewed the translation, and after some very constructive exchange, a final form was found.

Summer Morning (Marie von Ebner-Eschenbach)

On mountain peaks bedecked with snow,
On green hills stretching wide below,
The morning sun is shining.
Young beechwoods raise their dewy boughs
And tremble, tremble with the joys,
The very joys of living.

Down from the darkest rocky height
In total unrestrained delight
The roaring torrent rushes.
Its breath wakes life below, above,
In noble tree and lowly shrub,
And in the yielding mosses.

And over where the meadows lie,
The swarms of bees and midges fly
And hum in floral splendour.
See them in high grass stirring there
In busy movement full of cheer,
Buzzing with voices tender.

The young lark rises up so free,
Soars like a shout of jubilee,
A-whirling and a-singing.
In woods nearby the cuckoo calls,
And through the air the song-thrush sails,
On golden plumage winging.

Bright sunlit world! Delectable
Today; the rousing story;
Replete and lofty treasuries!
And yet, alas – ephemeral,
And doomed to death, for nature is
Sheer anguish, cloaked in glory.

The photo on the next page is a beautiful visualisation of this poem.

Even such small, incidental discoveries on a winter afternoon at the beach by the Baltic Sea, where this bird feather with ice crystals and first drops of molten ice show the complex structures in our world and touch us with their beauty.[4] At the same time, they point to some secrets: How did it come about that the feather is here, that these diverse and complex ice crystals have grown (and nowhere else in the immediate vicinity), and much more if we delve into the details of this photo. At the same time, we see the transitoriness, disguised in beauty; or, as the poem says in poetic English: "ephemeral, doomed to death, cloaked in glory". (Photo by the author).

---

[4]You may have sensed while reading this book that I am constantly looking around, eager to observe and admire the complexity and beauty in the world in which we are allowed to live. That's why I have perceived all of my life to be very rich and meaningful. You can find some more objective confirmation of my subjective experience here: https://www.scientificamerican.com/article/a-new-dimension-to-a-meaningful-life1/

Appendix

The following chapters of the appendix will also be easy to understand. I don't intend to make you experts on the various issues with these texts. But we'll go a little deeper than in the book. I refer those of you who want to go even deeper and actually learn the scientific details to the numerous literature references. There, you can find scientific articles by many authors including myself; especially, of course, by the leading non-equilibrium thermodynamicists such as Ilya Prigogine, Grégoire Nicolis, Werner Ebeling and Rainer Feistel, from whose publications I have also quoted.

You will find the appendix on the Internet at https://doi.org/10.1007/978-3-658-37755-7_1.

Table of Contents of the Appendix

Printed in the United States
by Baker & Taylor Publisher Services